Thin Films - Fundamentals, Deposition Techniques and Applications

Edited by Pablo Cayado

Published in London, United Kingdom

Thin Films - Fundamentals, Deposition Techniques and Applications
http://dx.doi.org/10.5772/intechopen.1004509
Edited by Pablo Cayado

Contributors
Abdelhakim Bouissil, Amal Hammoud, Badriyah Alhalaili, Celal Yelgel, Djallel Eddine Touaibia, Frederic Sanchette, Mohamed El Garah, Mpfunzeni Raphulu, Nafees Ahmad, Najoua Kamoun-Turki, Ruxandra Vidu, Siphelo Ngqoloda, Sofiane Achache, Thelma Ngwenya, Övgü C. Yelgel

First published in London, United Kingdom, 2025 by IntechOpen
IntechOpen is the global imprint of INTECHOPEN LIMITED, registered in England and Wales, registration number: 11086078, 167-169 Great Portland Street, London, W1W 5PF, United Kingdom

For EU product safety concerns: IN TECH d.o.o., Prolaz Marije Krucifikse Kozulić 3, 51000 Rijeka, Croatia, info@intechopen.com or visit our website at intechopen.com.

British Library Cataloguing-in-Publication Data
A catalogue record for this book is available from the British Library

Thin Films - Fundamentals, Deposition Techniques and Applications
Edited by Pablo Cayado
p. cm.

This title is part of the Materials Science Book Series, Volume 9
Topic: Surface Science
Series Editor: Chonghe Li
Topic Editor: Wan Sharuzi Wan Harun

Print ISBN 978-1-83634-323-3
Online ISBN 978-1-83634-322-6
eBook (PDF) ISBN 978-1-83634-324-0
ISSN 3049-8856

IntechOpen Book Series
Materials Science
Volume 9

Aims and Scope of the Series

Materials science has always occupied an extremely high position in the human development process. As we explore the oceans of stars, various industries have put forward more stringent requirements for the performance of materials, forcing us to pay more and more attention to the development of new materials. At the same time, the formation of a data-driven scientific paradigm is dramatically shortening the development cycle of new materials. The huge data generated by synergistically combining theories, high-throughput experiments, high-throughput computation, and artificial intelligence is greatly contributing to our ability to utilize materials science to solve real-world problems. The three topics of this book series - Metals and Nonmetals; Composite Materials; and Surface Science - will address important areas of advancement in materials science. There will be a range of interesting works published under these topics.

Meet the Series Editor

Prof. Chonghe Li received his Ph.D. from the Chinese Academy of Sciences in 1995. From 1995 to 2000, he worked as a researcher at the Shanghai Institute of Metallurgy, Chinese Academy of Sciences, where he also served as director of the research laboratory. In 2000, he was appointed Professor at the Institute of High-Performance Computing in Singapore, where he worked on computation and simulation in materials science until 2004. Since then, he has been a professor at the School of Materials Science and Engineering, Shanghai University, China, as well as the director of the Shanghai Specialty Casting Engineering and Technology Research Center. Prof. Li's research focuses on titanium alloys, titanium-aluminum single crystals, intermetallic compounds, theoretical calculations, alloy design, and special refractory materials. His broad scientific expertise is well recognized by the scientific community around the world. He is a member of the editorial board of the journal Metals. As an author, he has published more than 200 peer-reviewed papers, 2 books, and over 40 patents.

Meet the Volume Editor

Pablo Cayado (ORCID: 0000-0003-3703-6122) is a Lecturer at University of Oviedo, Spain, since January 2025. Previously, he was a Research Associate at the Department of Quantum Matter Physics (DQMP) at the University of Geneva in Switzerland for 4 years, and before that, he worked for another 4 years at the Institute of Technical Physics at the Karlsruhe Institute of Technology in Germany. He studied physics at the University of Oviedo and graduated in 2011. Afterwards, he moved to the Material Science Institute of Barcelona (ICMAB), where he completed his master's degree in Nanotechnology and Materials Science (2012) and Ph.D. in Materials Science (2016), working on the preparation and characterization of REBCO nanocomposite films by chemical solution deposition and the oxygen diffusion in these films. His current research activities focus on the electrical and mechanical properties of coated conductors for the development of REBCO superconducting joints. He is also working on using chemical solution deposition for REBCO films with new pinning scenarios (mixing of different REBCO compounds and new nanoparticles) and on the optimization of the oxygenation process in such films.

Contents

Preface

Thin films play a crucial role in modern science and technology and are now essential for the development of many key technological fields. With thicknesses ranging from just a few nanometers to several micrometers and made from a wide variety of materials, thin films are now fundamental components in numerous applications. Their importance lies in their unique properties, which often result from complex physical phenomena that emerge when materials are reduced to tiny dimensions. These properties can also be precisely tailored at the nanoscale using advanced deposition methods and material design techniques, making thin films a subject of intense research across many disciplines.

The continuous progress in thin film technology requires an integrated perspective that connects fundamental knowledge with process development and practical use. This book aims to support that approach by offering a broad review of the fundamentals, deposition techniques, and applications of thin films. The five chapters in this volume offer both general overviews and in-depth discussions of key topics, including high-voltage electronics, transition metal oxides for photocatalysis, high-entropy thin films, solar cell technologies, and the spray pyrolysis method for emerging materials.

The book is organized into two main sections. The first section focuses on deposition techniques and the optimization of thin film fabrication, while the second highlights advanced applications of thin films. The chapters cover a range of relevant subjects, including the role of thin films in high-voltage and power electronics; the design and functionalization of transition metal oxides for photocatalytic use; the fabrication of high-entropy thin films with improved resistance to oxidation; the development and characterization of promising materials like Cu_2MgSnS_4 for low-cost solar cells; and recent advances in the deposition of thin films for photovoltaic applications.

This book is intended to serve as a valuable reference for researchers, engineers, and students in the field of materials science. It provides an up-to-date and well-structured overview of current developments that will be of interest to a broad scientific audience.

I would like to thank the contributing authors for their excellent chapters and valuable input. Their expertise and dedication were essential to the creation of this volume.

Pablo Cayado
Department of Physics,
University of Oviedo,
Oviedo, Spain

Section 1

Deposition Techniques and Process Optimization

Chapter 1

High-Entropy Films by Magnetron Sputtering: Deposition Parameters, Oxidation Resistance Enhancement, and Applications

Djallel Eddine Touaibia, Abdelhakim Bouissil, Sofiane Achache, Frederic Sanchette and Mohamed El Garah

Abstract

Coatings play a key role in modern industry, enhancing the performance of materials. The diversity of their applications and constant progress make them a central area of research and development in materials science and engineering. In this manuscript, we examined some recent works of high-entropy coatings deposited by magnetron sputtering. The first section provides details on the magnetron sputtering technique and deposition mechanism. The change of the parameter influences the microstructure and then the properties of the films. High Power Impulse Magnetron Sputtering is sued to increase the compactness of the film. In the second, a spotlight on High-Entropy Films (HEFs) as an emergent-class material is presented and how their oxidation resistance is improved. Particular attention is being paid to studying the effect of some alloying elements, such as nitrogen and silicon, on oxidation resistance improvement. The last section presents potential applications of these coatings, especially the cutting tools, the diffusion barrier, and other industrial uses.

Keywords: coatings, PVD, high-entropy films, oxidation, temperature

1. Introduction

Coating is the process of applying a thin layer of material to a substrate surface by either liquid phase (solution) or solid phase (powder or nanoparticles). Coating techniques can be adjusted to meet production requirements for coating thickness, surface roughness, rate, and product size, which are determined by velocity, film width, and patterning capabilities [1–3]. Deposition of a layer (ranging from a few nanometers to hundreds of microns) on a surface is now widely employed for a variety of industrial purposes. The primary objectives of thin layer coating are to improve the surface's effectiveness, service life, and quality. The coating is frequently used in industry due to its practical features and applicability.

The coating process can be carried out under either atmospheric pressure or low vacuum conditions. It is classified according to the solubility of the coating materials being considered. In solution coating, a binder and target material are dissolved in a solution that can directly cover the substrate, and the solution evaporates from the wet film to form a dry film. In contrast, vacuum deposition techniques include chemical vapor deposition (CVD) [4], physical vapor deposition (PVD) [5], atomic layer deposition (ALD) [6], plasma [7], and sputtering [8]. The latter technique is the most widely used, and produces dense, well-adhered films and homogeneous chemical composition of the deposited coatings. This technique offers the possibility of fine-tuning the properties of the deposited films [9]. Magnetron sputtering is a physical vapor deposition technique based on the principle of condensing metal vapor from a target onto the surface of a given substrate.

Providing coatings with higher performance is of foremost importance for many applications. Therefore, since their emergence in 2004, much work has been done on high-energy alloys (HEAs) [10–13]. These latter can be divided into seven families: 3d transition metal alloys, refractory metal alloys, light metal alloys, lanthanide (4f) transition metal alloys, brasses and bronzes, precious metal alloys and interstitial compound (boride, carbide, and nitride) alloys [14]. High-entropy oxides can also be added to these families, which are particularly interesting in the field of superconductivity.

Various techniques have then been implemented to deposit HEFs [12]. Metallic and ceramic HEFs are being extensively researched; nitrides and carbides, for example, are attractive functional materials. The interest lies in the good physicochemical properties and the wide range of material design possibilities. However, these improvements are accompanied by significant problems in terms of composition selection. Traditional materials science methods may be slow to improve. To overcome these problems, a new method is required. As a result, information technology has the potential to provide materials science with incredible tools to make significant progress [15].

HEAs are composed of at least five elements with atomic content varying from 5–35% [16, 17]. Single-phased solid solutions, free of intermetallic phases, are reported as the most investigated alloys. HEFs, which are inexpensive, have also been studied [11, 18, 19] and have shown interesting physical and chemical properties. The characteristic keys of HEAs and HEFs are presented below:

- High configurational entropy: The random arrangement of multiple elements in equimolar quantities creates a highly disordered crystal structure, resulting in improved mechanical and thermal properties.
- HEAs have excellent mechanical properties, including high strength, hardness, and ductility, making them suitable for various technical applications [20–22].
- High-entropy composition stabilizes solid solutions and inhibits phase transitions at high temperatures.

HEFs are used in variety of applications such as electrical devices, and sensors where thin and functional materials are required.

On the other hand, the key features of HEFs are presented below.

- Conformal coating: HEFs can be deposited as thin, conformal films on a variety of substrates, providing uniform coverage even on complex geometries, such as printed circuit boards.

- Tailorable features: Because the composition contains a variety of materials, researchers can tailor the film's properties for a specific application.

- Enhanced surface qualities: The nano- and micro-scale thickness of HEFs results in different surface properties and interactions, making them extremely useful for surface engineering [23].

- Due to their superior properties, high-entropy coatings can be considered as alternatives for high hardness coatings, thermal barrier coatings for turbine blades, advanced nuclear applications, and diffusion barriers in integrated circuit (IC) systems [24].

2. Magnetron sputtering

Magnetron sputtering is a physical vapor deposition (PVD) process that is widely used in the manufacture of coatings. It is characterized by its efficiency and control of film properties through a variety of deposition parameters. This technique is essential in industry and academic laboratories, particularly for surface engineering, because of its ability to produce uniform, high-quality coatings. Magnetron sputtering can be performed in a variety of configurations, including direct current (DC), radio frequency (RF) and pulsed DC. Each mode is tailored to the type of material and the desired coatings characteristics.

The synthesis of coatings by magnetron sputtering can be done in three steps as shown in **Figure 1**: (1) generation of metal vapor by extracting atoms from a solid (negatively polarized) target, (2) transfer of the vapor to the substrate in a gaseous atmosphere, and (3) condensation of the vapor on the substrate followed by germination and growth of the coating [8, 25].

- Metal vapor generation (sputtering): Target (or source; material to be deposited) and substrates are placed into a vacuum chamber. When a high voltage of electricity is applied, positively charged argon ions (Ar^+) are attracted by the target,

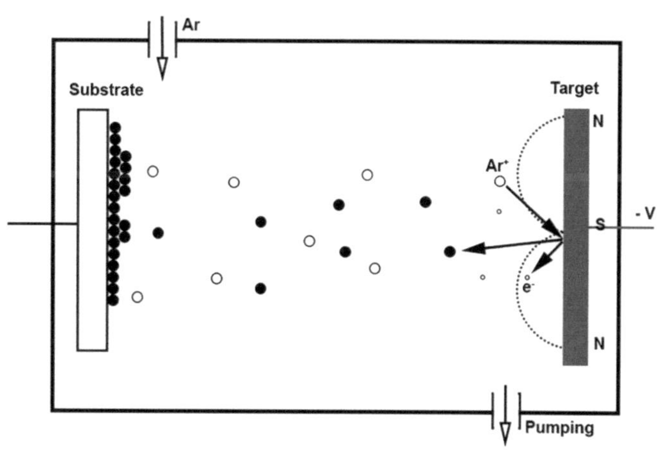

Figure 1.
Magnetron sputtering principle.

which acts as cathode. This luminous electrical discharge is self-sustaining, in part due to the electronic secondary emission from the target elements. It is a cold plasma, with a metal vapor ionization rate of a few percent. Ar^+ ions are accelerated toward the target by the electrostatic field and can cause four phenomena described in **Figure 1**. The Ar^+ ions collide with the target atoms, transferring energy and ejecting one or more atoms from the surface. The electrical discharge is sustained by secondary electronic emission.

- Metal vapor transfer to the substrate: Atoms sputtered from the target surface have an average energy of about 5 to 10 eV. During the transfer of metal vapor atoms from the target to the substrate, they collide with argon atoms, resulting in a loss of kinetic energy. The number of collisions is related to the distance traveled by the sputtered atom and the operating pressure and determines the likelihood of shocks in the plasma. For example, for an argon discharge at 1 Pa, the distance between two successive collisions with argon atoms (Free Mean Path, FMP) is approximately one centimeter [8]. Note that the FMP is also affected by the species in the plasma phase. Since the small distance (few centimeters) between the target and the substrate to be coated, the sputtered atoms lose most of their chemical energy and thus bind to the substrate in a thermalized state with a chemical energy of about 0.1 eV. Reducing the working pressure allows the average free path to increase while maintaining an adequate energy level (for surface diffusion) of the atoms arriving on the substrate [8, 25].

- Condensation and growth on the substrate: The energy of the sputtered atoms plays a key role in the adsorption mechanism. If the atoms arrive at the substrate with low energy, adsorption will take place at the arrival sites. However, if the atoms' energy is higher, they can diffuse across the surface of the growing film to more stable sites, improving the compactness. **Figure 2** shows the surface diffusion process and the adsorption of atoms onto the substrate to be coated. The energy required for atomic surface diffusion can be supplied directly during sputtering (discharge parameter) or by limiting the loss of energy during transport (work at reduced pressure). Another way of supplying energy is to add a substrate heating system or a DC or radio frequency (RF) power supply system [25, 26].

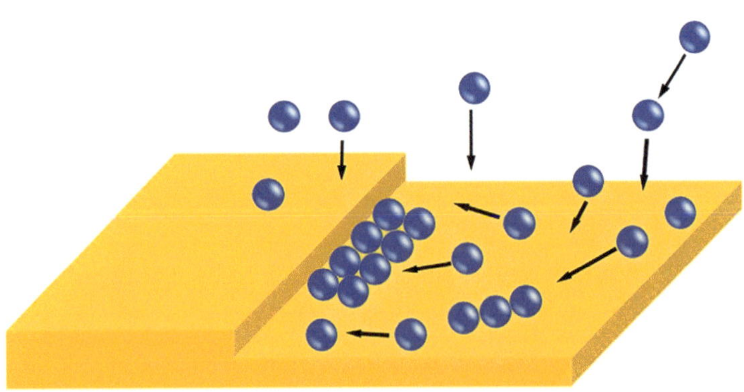

Figure 2.
Illustration of atoms' adsorption and coating growth onto the substrates.

2.1 Operating modes in magnetron sputtering

2.1.1 Direct current magnetron sputtering (DCMS)

The DC magnetron sputtering process uses of a DC power supply to generate a plasma in a vacuum chamber. Inside the vacuum chamber, the target (cathode) is bombarded with energetic ions, usually from an inert gas, often argon. The ions finally dislodge the atoms from the target leading to their sputtering. DC magnetron sputtering is particularly well suited to the deposition of conductive materials because the use of DC current ensures stable plasma conditions. However, difficulties arise with insulating materials due to the accumulation of positive charges on the target surface. This leads to arcing and process instability. DC sputtering is then a simple technique in which a voltage is applied between the target and substrate. Power, bias voltage, and argon pressure can be adjusted to control the deposition rate [26, 27].

2.1.2 Radio frequency (RF) magnetron sputtering

In contrast to DC sputtering, radio frequency (RF) process is used to deposit insulating materials. With DC sputtering, a very high voltage of the order of 1012 V is required to deposit an insulating film. This can be avoided with RF sputtering. RF sputtering allows lower argon pressure in the plasma, resulting in fewer gas collisions and more line-of-sight deposition [26, 27]. The plasma of a 'normal' magnetron discharge, whether direct current (DC), medium frequency (MF) pulsed, or radio frequency (RF), is mainly a gas plasma. The most common process gas is argon or, in the case of reactive sputtering, a reactive gas such as nitrogen or oxygen. Unless the pressure is extremely high, the sputtered atoms move neutrally through the plasma gas to the substrate without colliding. It is important to note that most of the sputtered atoms remain neutral, and the films are formed by the condensation of the atoms [28].

2.1.3 High power impulse magnetron sputtering (HiPIMS)

HiPIMS is a highly ionized physical vapor deposition process. It produces dense, smooth, and well-adhered coatings. Unlike other conventional sputtering methods, where pulses are delivered at modest power and at a repetition rate of several kHz, HiPIMS technology uses high-power pulses that are two orders of magnitude higher than normal magnetron power. In HiPIMS technology, the high-power discharge pulses result in a high degree of ionization of the atoms (several tens of percent), which has important consequences for the microstructure of the film. In general, the HiPIMS operation can be performed using the same magnetrons as for standard sputtering: simply replace the magnetron power supply with a HiPIMS power. HiPIMS has the advantage of producing good quality films, but the low deposition rate seems to limit the market penetration of HiPIMS technology [28].

2.1.4 Reactive mode

Reactive sputtering involves sputtering an elemental target in the presence of a reactive gas, which reacts with its material to form a composite. This is the case for the formation of ceramics such as oxides, nitrides, and carbides when oxygen, nitrogen, or methane are added, respectively. Reactive sputtering was developed in the 1950s to deposit doped Ta-N films for integrated circuits, and its use has multiplied since. The gas reacts with the target material due to the presence of the high voltage of electricity; other reactions with the chamber wall may also occur. In this mode, a composite is formed on the surface of the target. The deposition rate therefore decreases as the surface becomes covered with this composite, and once completely contaminated, the target is poisoned.

2.2 Effect of parameters deposition on films properties

In magnetron sputtering, several parameters affect the growth of the film and, subsequently, its properties. At high power, an increase in elemental sputtering can occur. The HiPIMS approach achieves high ionization in sputtered materials by pulsing the target with a high peak power density (1–3 kW/cm^2). This allows fine-tuning of ion bombardment intensity and flux to the substrate, resulting in superior control of film growth [29–31]. The HiPIMS deposited coating has a much-refined structure, indicating a higher film density and smoother surface structure compared to DC deposited coating. This is thought to be related to improved diffusivity of the atomic species on the surface, which help to fill the holes [32, 33]. In addition, the deposition rate in HiPIMS mode was measured at 12.92 nm/min, which is lower than that of DC mode (17.28 nm/min). This decreasing trend of the deposition rate could be explained by increased re-sputtering at the growing film surface. This latter is caused by the increased energy of the ions in the plasma [32, 33]. Also, HiPIMS technology can help to improve the hardness [32].

The working pressure of the gas can influence the mean free path of the sputtered elements. It is inversely proportional to this pressure. When the gas pressure increases, the mean free path becomes shorter, and the sputtered elements undergo more collisions before being deposited on the substrate surface. These collisions result in a low deposition velocity, leading to low atom mobility at the substrate surface. Finally, the films obtained will have a low compactness, which may affect their properties. Temperature can also be used to control the properties of films. Increasing the temperature leads to greater mobility of the atoms on the surface of the substrate. This increases the grain size. It may also stabilize a phase that cannot be achieved by deposition at room temperature. However, this temperature must not be exceeded to avoid damaging the films [34]. In addition, the substrate bias applied during the sputtering process can significantly influence the physical and chemical properties of the films. The adhesion of the film to the substrate can be improved. Under these conditions, a more compact and dense film can grow due to the mobility of the adatoms. Polarization of the substrate can prevent the growth of large grains due to the impact of high energy ions on the film.

2.3 Thin films growth

The sputtering process involves the condensation of vapor phase elements on the substrate, resulting in the growth of a thin film. As with all phase transitions, this condensation is characterized by the nucleation and subsequent growth of nuclei. Depending on the interaction energies (affinities) between the atoms of the substrate and those of the deposited layer, three growth modes can be distinguished (**Figure 3**).

- In the two-dimensional Frank van der Merwe mode, the film grows layer by layer. The affinities between substrate and film atoms are higher than those between adjacent film atoms.

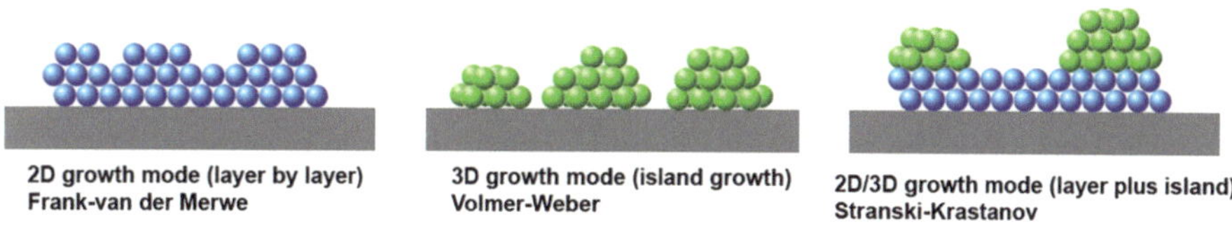

Figure 3.
Growth modes of thin films.

- In the Volmer-Weber mode, separate three-dimensional islands are formed on the substrate. The interactions between the film atoms are greater than those between the film and substrate atoms.

- The Stranski-Karastanov mode is characterized by layer plus island. One or two monolayers are formed first, followed by germination of separate islands.

Growth is the stage that follows nucleation. Whereas nucleation is influenced by the interactions between substrate and film atoms. It is governed by the kinetic energy of the sputtered ad-particles. This energy is the driving force behind the mobility of the species at the surface of the growing film.

The growth of high-entropy films using the magnetron sputtering technique involves a complex interaction of nucleation, atomic mobility, and competition between different elements. Due to high configurational entropy, the formation of a solid solution can be reached. The target is typically a combination of multiple targets containing different elements. When the target is bombarded with ions, all atoms are sputtered and deposited onto the substrate. This results in a flux of multi-element atoms reaching the substrate. These atoms undergo nucleation, depending on different factors such as mobility of atoms on the surface, substrate temperature, etc. The atoms initially form small clusters. Besides, if the surface mobility is higher, uniform crystalline structure can be obtained. The competition between different elements, constituting the HEF, during nucleation and growth, helps to avoid phase segregation and encourages the formation of a solid-solution phase. Consequently, a single-phase structure such as body-centered cubic (BCC), face-centered cubic (FCC), or hexagonal close-packed (HCP) can be formed.

3. High-entropy films

As mentioned above, considerable work has been carried out on the deposition of HEFs [12]. Their ultimate properties lie in the combination of processing, composition and microstructure [35, 36]. They are often composed of hafnium (Hf), molybdenum (Mo), tungsten (W), zirconium (Zr), vanadium (V), and niobium (Nb). The addition of elements such as aluminum (Al) and silicon (Si) can improve the microstructure, mechanical properties, and oxidation resistance of HEFs [36, 37]. Based on the previously established classification system [14, 38, 39] and the state-of-the-art for HEAs and their related coating materials. HEFs can be divided into three categories:

- Metallic HEFs, including transition and refractory elements: transition metal HEFs typically contain elements such as Co, Cu, Fe, Mn, and Ni. An example is the Cantor alloy CrMnFeCoNi [40, 41]. RHEFs are mainly composed of elements with high melting temperatures (Tm) such as Hf (2200°C), Mo (2623°C), Nb (2477°C), Ta (3020°C), Ti, V, W, and Zr, with non-refractory elements such as Al (660°C) being connected [14]. Metallic HEFs tend to resist abrasion and can help to protect the coated substrate from corrosion. The presence of refractory materials often reduces oxidation at high temperatures [39, 42].

- HEA-based ceramic coatings, such as those produced by combining HEAs with oxygen, boron, and other anions, form covalently and ionically bonded layers [14]. Ceramic HEFs are mainly based on transition metals, which can easily form binary nitrides (TiN, TaN, HfN [43]) or binary carbides NbC, CrC, and WC. Various nitrides, carbides, oxides, and carbonitrides can be deposited using reactive deposition techniques [44]. Ceramic HEFs with appropriate compositions and processing parameters often outperform conventional binary nitrides, carbides, oxides and carbonitrides, mainly due to the combined effect of high entropy, lattice distortion and slow diffusion [35, 45, 46]. Several nitride HEFs have been prepared by reactive DCMS, such as TiVCrAlZr(N) [47] and AlCrMoTaTiZr(N) [48].

- Composite HEFs, in which HEAs act as the matrix/binder with the hard ceramic reinforcements or synthesized as the hard reinforcements like TiN [49], NbC [50], TiC, or TiB_2 [51], or synthesized as the hard reinforcements in lightweight alloys such as Al and Mg alloys. Composite HEFs with excellent properties have been developed, including high hardness, high wear resistance, and chemical and thermal stability. These include TiN/$CoCr_2FeNiTi$, NbC/AlCoCrFeNi coatings [50] deposited by laser cladding, and TiC-TiB_2/CoCrCuFeNi layers synthesized by transferred-arc plasma [51] can be listed.

3.1 High-temperature oxidation of HEAs

To advance the use of HEAs in real-world applications, it is necessary to examine their oxidation behavior under various situations. The oxidation of an alloy is a chemical reaction between oxygen and its constituents. The formation of oxides can be both beneficial and detrimental to the functional properties of materials, thereby limiting their applications. If the oxide formed is stable, it can act as a barrier layer to the diffusion of oxygen. For example, aluminum and chromium can form protective layers. However, this is still not the case for all materials. Most oxides form heterogeneously and are not protective. They therefore weaken the surfaces of materials and lead to considerable loss of mass, particularly at high temperatures. It is therefore very interesting to improve the oxidation resistance of materials.

3.1.1 Nitrogen effect

The study of high-temperature oxidation is of particular interest in terms of alloy utilization. The study of HEA oxidation is useful for understanding coating production and high-temperature gas corrosion behavior. The potential for the use of HEAs in high-temperature environments is great; however, understanding their oxidation behavior is quite complex [52]. The oxidation of HEAs leads usually to the volume expansion of the material and the degradation of its properties, especially the mechanical properties. Nevertheless, due to their unique structure and properties, high-entropy alloys are promising candidates for anti-oxidation properties at high temperatures. In this section we focus on the oxidation resistance of HEA coatings. Recently, HEAs with improved high-temperature oxidation resistance have been developed [53]. Touaibia et al. [24] synthesized $(AlTiZrHfTa)_{1-x}N_x$ refractory HEFs by direct current magnetron sputtering under various $R_{N2} = N_2/(Ar + N_2)$ ratios, and depicted the nitrogen effect on oxidation resistance of the coating at 800°C. The introduction of nitrogen into the film significantly reduced the kinetic constant k_p from 1.77 10^{-6} g^2 cm^{-4} h^{-1} to 1.22 10^{-7} g^2 cm^{-4} h + at 800°C. Furthermore, **Figure 4** clearly shows that in the case of AlTiZrHfTa nitrogen-free film, the oxidation is complete, resulting in a cracked and porous morphology. However, the nitride films obtained at R_{N2} = 10, 20, and 40% are not completely oxidized, revealing a better oxidation resistance compared to that of the nitrogen-free film [24].

In fact, according to thermogravimetric analysis (TGA) results (**Figure 5**), a beneficial effect of nitrogen addition on the oxidation resistance of films at 700, 750, and 800°C can be observed. By adding nitrogen to the films, ionic-covalent bonds are formed, which could lead to improved oxidation resistance. A slower initial parabolic growth in mass gain is observed at 800°C, followed by a rapid linear increase (**Figure 5**). The decreasing of the gain mass values corresponds to the improved oxidation resistance, as can be clearly seen on curve of metallic film (sky blue color) and the nitrides (black, green, and red color). The linear increase should correspond to a loss of the protective character of the oxide layer. However, in the case of nitrides, no catastrophic oxidation is observed. A clear improvement in oxidation resistance is therefore observed for these nitrides compared to the metal film.

The $(AlTiZrHfTa)_{1-x}N_x$ nitride obtained at R_{N2} = 10% was analyzed by Transmission Electron Microscopy (TEM). The film, as deposited, shows columnar growth (**Figure 6a**) with a preferential orientation (111), as shown by the electron diffraction pattern. After oxidation, the film shows two different zones: an intact film with black contrast at the bottom and a thick, porous, homogeneous oxide layer at

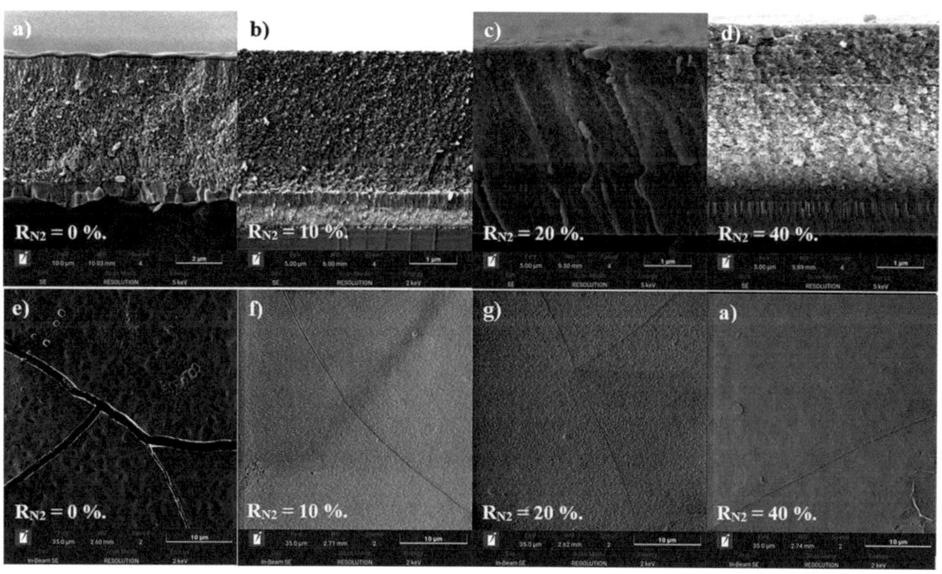

Figure 4.
Cross-sectional and surface SEM micrographs of $(AlTiZrHfTa)_{1-x}N_x$ RHEFs for R_{N2} = 0, 10, 20, and 40% after 1-hour oxidation at 800°C in the air. The figure is reproduced from open access reference [35].

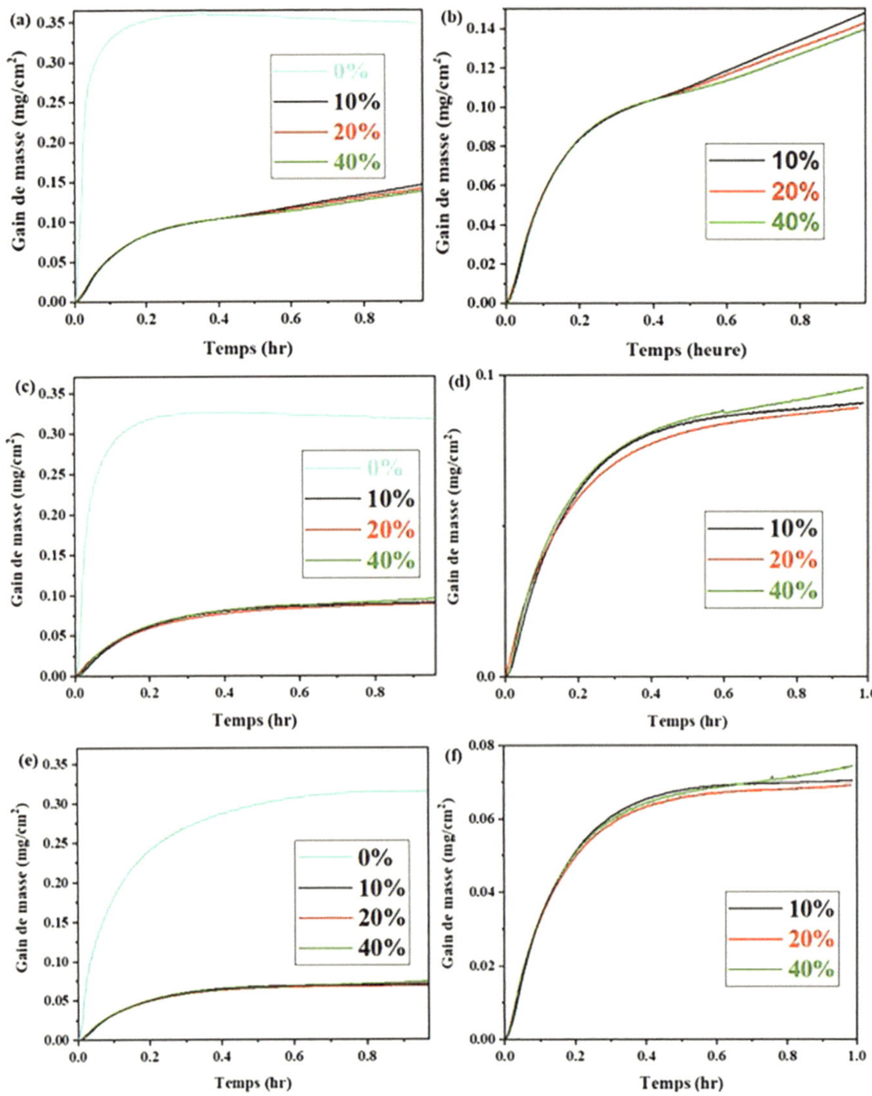

Figure 5.
Variation in mass gain per unit area as a function of time for (AlTiZrHfTa) and $(AlTiZrHfTa)_{1-x}N_x$ films at 700, 750, and 800°C, respectively. The figure is reproduced from open access reference [35].

the top (**Figure 6b**). A zoom of the oxidized film (**Figure 6c**) shows a porous layer, confirming the hypothesis put forward in the previous paragraph.

The high-resolution micrograph (HRTEM) of the oxide layer is shown in **Figure 6d**. Two phases, crystalline and amorphous, coexist. This is also confirmed by the Fast Fourier Transform (FFT), which reveals bright spots and a diffuse circular ring corresponding to the crystalline and amorphous phases, respectively. The size of these crystalline domains is nanometric, making them difficult to identify.

For further analysis, STEM-EDS mapping of the film before and after oxidation was carried out (**Figure 7**). The results show a homogeneous distribution of metallic elements in the nitride. After 1 hour of oxidation at 800°C, a homogeneous distribution of metal elements in the oxide layer is also observed, revealing the absence of

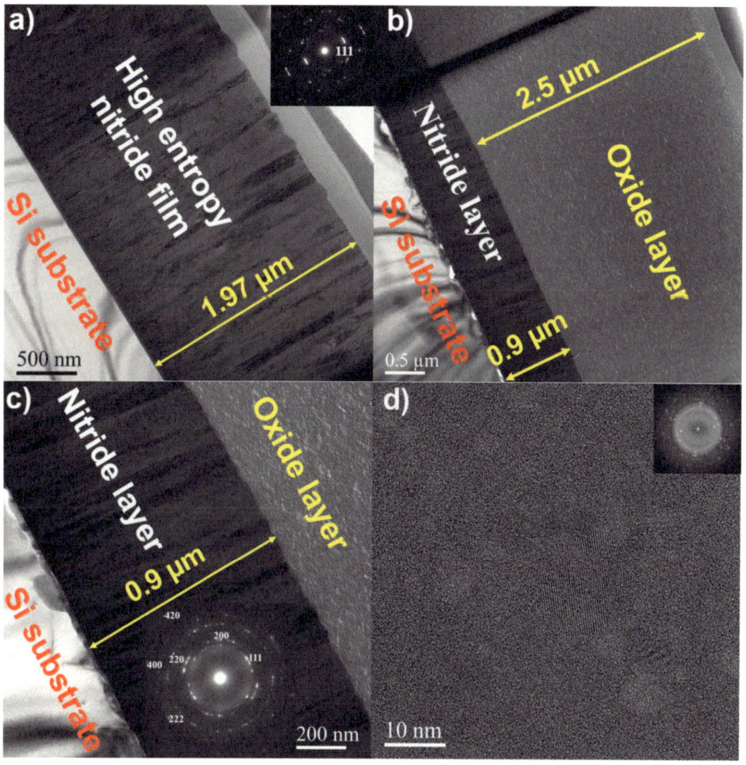

Figure 6.
Bright-field TEM micrograph of nitride $(AlTiZrHfTa)_{1-x}N_x$ obtained for R_{N_2} = 10%, and the associated electron diffraction pattern (b) bright field TEM micrograph of the same film after oxidation in air at 800°C. (c) Zoom in on the non-oxidized zone of the $(AlTiZrHfTa)_{1-x}N_x$ film obtained for R_{N_2} = 10% and the associated electron diffraction pattern (d) HRTEM image and corresponding Fast Fourier Transform (FFT) of the oxidized zone of the $(AlTiZrHfTa)_{1-x}N_x$ film obtained for R_{N_2} = 10%. Figure and adapted with permission from open access reference [35].

any preferential oxide. A mixed complex oxide appears to have formed. In addition, nitrogen mapping after oxidation clearly shows a decrease in nitrogen content throughout the oxide layer formed. This may point to nitrogen depletion in high-entropy nitrides during oxidation. This depletion would contribute to pore formation during oxidation.

Contrarily to the kinetic constant k_p, the activation energy E_a values evolved in the opposite direction, indicating an improvement in oxidation resistance due to the creation of strong ionic-covalent M—N bonds [24]. Bouissil et al. [54] observed the same tendency when investigating TiTaZrHfW(-N) coatings.

3.1.2 Incorporation of silicon into the films

Much research has been conducted in the past to improve the oxidation resistance of high-entropy coatings via alloying additions containing metals such as Al, Cr, and Si. When introduced in sufficient amount, these elements can promote the formation of protective oxide top-layers, such as Al_2O_3, Cr_2O_3, and SiO_2 [37, 55–58]. The deposition of the nanolayered FCC $(AlTiZrHfTa)_{Nx}$/a-SiN_x coating, by DCMS, improved the oxidation resistance at 800°C. This oxidation resistance improvement is illustrated by

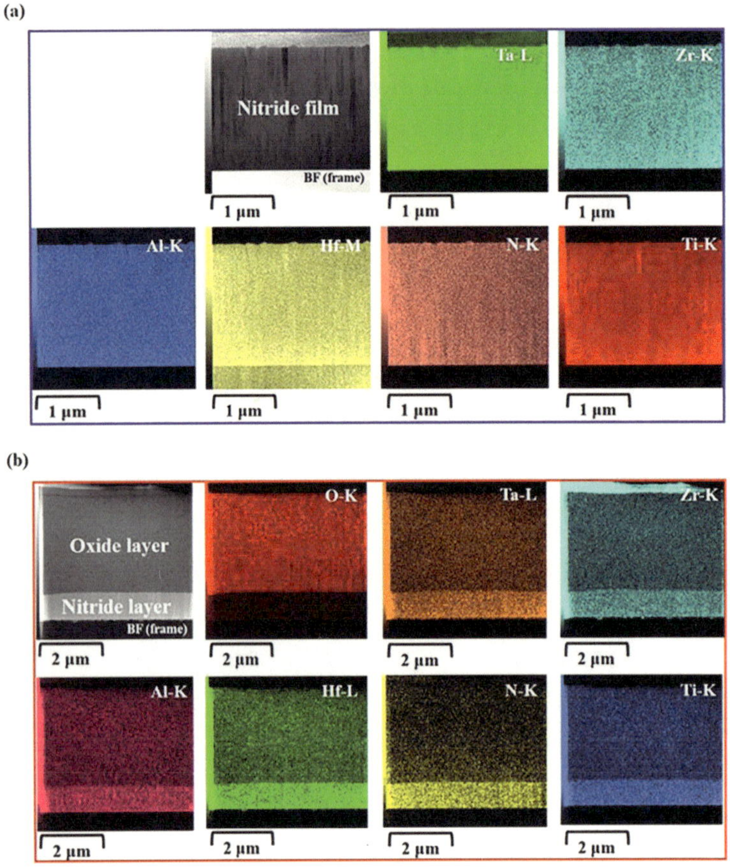

Figure 7.
STEM-EDS cross-section mapping of nitride $(AlTiZrHfTa)_{1-x}N_x$, obtained for R_{N2} = 10% (a) before and (b) after 1 hour's oxidation at 800°C. Figure reproduced with permission from open access reference [35].

the decrease of the oxide thickness from 2.5 to 0.6 μm for FCC $(AlTiZrHfTa)N_x$/a-SiN_x coating (**Figure 8**).

In addition, the activation energy E_a increased from 90.8 kJ·mol^{-1} for the FCC $(AlTiZrHfTa)N_x$ coating to 126.52 kJ·mol^{-1} for the FCC $(AlTiZrHfTa)N_x$/a-SiN_x nanolayer at I_{Si} = 0.2 A (I_{Si}: applied current on Si target). This trend reflects an improvement in oxidation resistance caused by the creation of the amorphous SiN_x nanolayer in alternative to FCC $(AlTiZrHfTa)N_x$ [37].

Bouissil et al. [59] also investigated $(WTiZrHfTa)N_x/SiN_x$ nanolayer by using magnetron sputtering technique. **Figure 9** presents TEM micrographs bright field (BF) and dark-field (DF) electron diffraction image (SAED) of the resulting architecture. The films are deposited with 9.3 at.%. The SAED diffraction pattern, characterized by continuous rings, confirms the presence of a nanocrystalline structure in the coatings, in line with what was mentioned in the section on X-ray diffraction. In addition, the DF-mode micrograph (**Figure 9e**) highlights the nanometric size of the grains. Increasing the amount silicon in $(WTiZrHfTa)N_x/SiN_x$ system (from 9.3 at.% to 17.7 Si at. % by increasing the Si target power) the results show a slight increase in the width of the SiN_x layers (brigh nanolayers). The width of the dark nanolayers appears to remain unchanged, which is consistent with no change HEA target

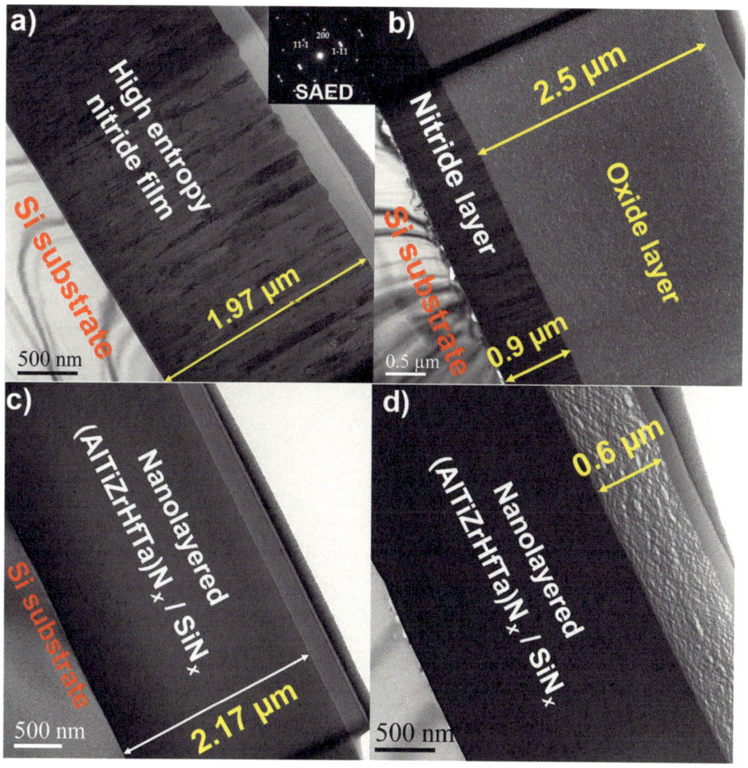

Figure 8.
(a) Bright-field TEM micrograph of intact nitride film at R_{N_2} = 10% with associated SAED pattern. (b) Bright-field TEM micrograph of the same film after 1-hour oxidation at 800°C. (c) Bright-field TEM micrograph of intact FCC (AlTiZrHfTa)N/a-SiNx film obtained for I_{Si} = 0.2 A with associated SAED pattern. (d) Bright-field TEM micrograph of the FCC (AlTiZrHfTa)N/a-SiNx film obtained for I_{Si} = 0.2 A after 1-hour oxidation at 800°C. Figure reproduced with permission from open access reference [37].

power. The light nanolayers are assumed to be silicon nitride, and the increase in their width could simply be due to the increase in power applied to the Si target during deposition.

The nanolayer architecture observed in the TEM is related to the deposition process. **Figure 10** shows schematically the growth of this particular architecture, resulting from the scanning mode during the deposition process. The two metal vapors land alternately on the samples, except when they are located between the two targets. The interface between the layers is therefore diffuse, as the vapor streams are homogeneously mixed in this zone. This may also be related to the fact that the distance between the targets and the substrate holder is only 6 cm, whereas the radius of the targets is 10.

During the swiping mode, the substrate holder moves through three zones (**Figure 10**). In the first zone, the substrates are exposed to the HEA target, forming a dark layer of TiTaZrHfW-N. In the second zone, the substrates encounter a mixture of two metal vapors, forming an interface layer with a gradient of silicon content. Finally, in the last zone (zone 3), the substrates are exposed only to the Si target, forming a clear layer of Si_3N_4 silicon nitride. In fact, at 600 and 700°C, the amorphous oxide layer could retard oxygen diffusion, whereas at 800°C, the oxide layer formed is nanocrystalline, and the grain boundaries could facilitate oxygen diffusion. This

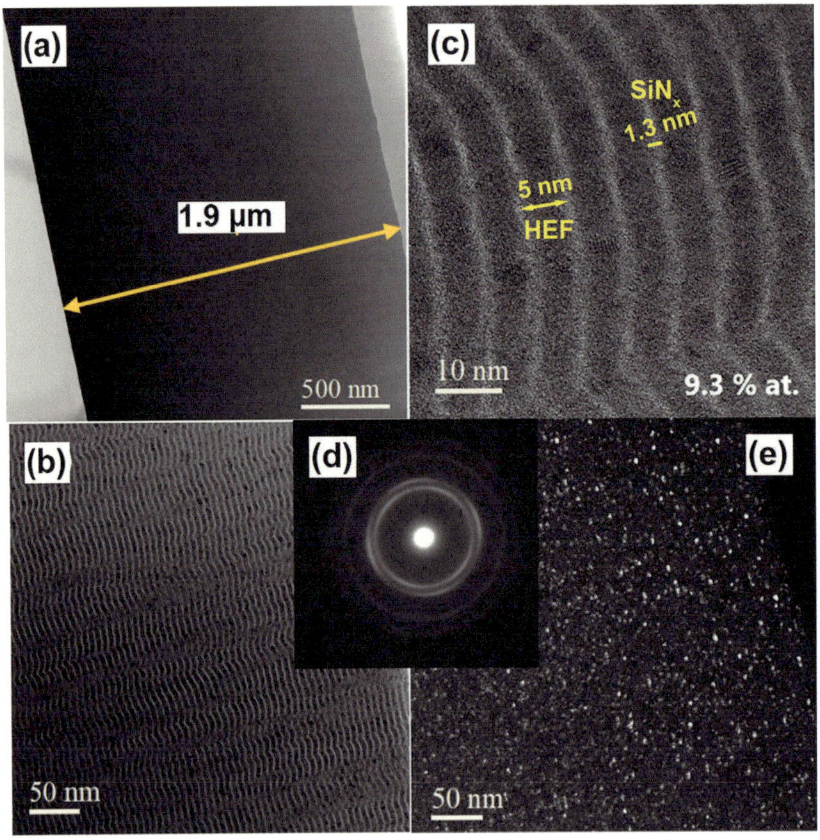

Figure 9.
(a) BF and DF (b) MET micrographs of (WTiZrHfTa)N_x/SiN_x cross-section. (c) Corresponding HRTEM micrograph, (d) corresponding SAE, and (e) HRTEM micrograph showing nanocrystals. Figure reproduced and adapted from the reference open access [59].

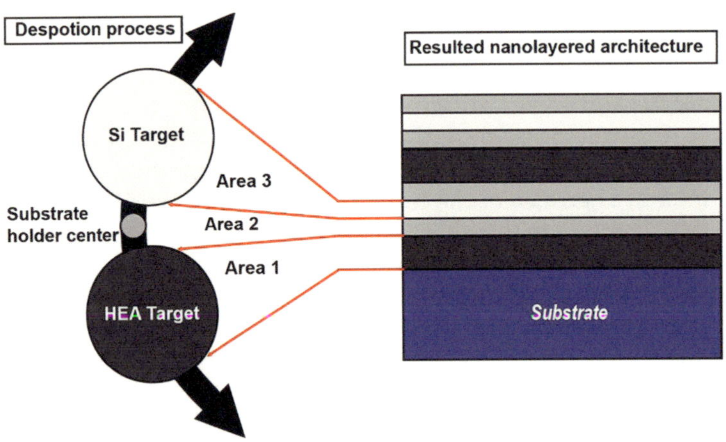

Figure 10.
Schematic showing the deposition procedure to prepare the nanolayered (AlTiZrHfTa)N_x/SiN_x or (WTiZrHfTa) N_x/SiN_x coatings.

could explain the TGA results, where a parabolic behavior of the mass gain has switched to a linear evolution when the temperature increases from 600 to 700°C.

The above results clearly indicate that the presence of SiN_x layer improvement is helpful in preventing oxidation. However, the mechanical properties of the HEFs are affected by their mechanical properties. This nanolayered architecture leads to the decrease in hardness and Young's modulus up to H = 17.7 ± 0.5 GPa and E = 162.5 ± 1.6 GPa FCC $(AlTiZrHfTa)N_x/a\text{-}SiN_x$ coating and up to H = 21 ± 0.3 GPa and E = 236 ± 2 GPa FCC $(WTiZrHfTa)N_x/SiN_x$ coating. The softening of the coatings results from the formation of the amorphous SiN_x nanolayers, hindering the growth of the FCC $(MTiZrHfTa)N_x$ (M = Al or W) nanolayers. Nanolayered architecture generally enhances the hardness of coatings by obstructing dislocation movements at interfaces [60]. TiN/Si_3N_4 multilayers, for example, showed an increase in hardness when the Si_3N_4 layer thickness was less than 1 nm (approximately 0.5 or 0.7 nm) [61, 62]. Dong et al. [63] found that the hardness of a ZrN/Si_3N_4 nanolayered coating changed as the thickness of the Si_3N_4 layer increased. Indeed, when the Si_3N_4 layer thickness is around 0.6 nm, the hardness increases due to the creation of a crystallized Si_3N_4 nanolayer, which forms coherent interfaces with the ZrN layer [37].

Furthermore, increasing the Si content in the films led to a slight improvement in oxidation resistance, which could simply be related to the increased thickness of the Si_3N_4 nanolayers (acting as a barrier to oxygen diffusion). The improvement in oxidation resistance could simply be due to both the multilayer architecture with amorphous Si_3N_4 nanolayers and even the formation of an amorphous oxide layer. This could act as a barrier to oxygen diffusion or at least slow down its diffusion. It can be assumed that the oxidation mechanism is anionic, resulting in the maintenance of a nanolayer architecture in the oxide layer. Nevertheless, increasing Si content reduces the slope of linear oxidation (especially at 800°C), which could be due to the increased thickness of Si_3N_4 nanolayers. However, increasing the thickness of Si_3N_4 nanolayers can reduce the mechanical properties of the film (low hardness). The challenge is to further improve oxidation resistance without degrading mechanical properties in the case of this film. Thus, the investigation of the oxidation resistance of the $(WTiZrHfTa)N_x/SiN_x$ has been performed only at 800°C (high temperature). TGA characterization was carried out with two different durations, 30 minutes and 10 hours, during oxidation at 800°C (not present here). A significant decrease in mass gain was observed, indicating an improvement of oxidation resistance. Even when compared with the film containing the highest amount of Si (17.7 at.%), these materials seem to have better resistance to oxidation. In addition, the linear nature of the mass gain curve evolved into a parabolic curve, which could be related to the formation of a more stable and protective oxide layer. As the oxidation resistance of these nanolayers was good for 30 minutes, the test time was then extended to 10 hours. The results show a better oxidation resistance of the sample with 17.7 at.% Si.

Kretschmer et al. [64] also investigated the oxidation behavior of (AlCrNbTaTi)N coatings in air as a function of Si content. The annealing was performed at 850°C for 100 hours. The Si-free coating had an oxide thickness of 2700 nm, while those with Si had 280 nm (**Figure 11**). The addition of Si to the coatings resulted in the creation of a thick microstructure, slowing down drastically the diffusion of oxygen. This could be due to the presence of SiO_2 amorphous, which improves the oxidation resistance [65].

Using cross-sectional SEM investigations of the coatings after 1, 10, 30, and 100 hours of oxidation at 850°C (with an additional 5-hour measurement for the Si-free coating), the evolution of oxide scale thickness was studied. This is exemplified by the Si-free and 6.4 at.% Si-containing samples after 100 hours (**Figure 11**). The

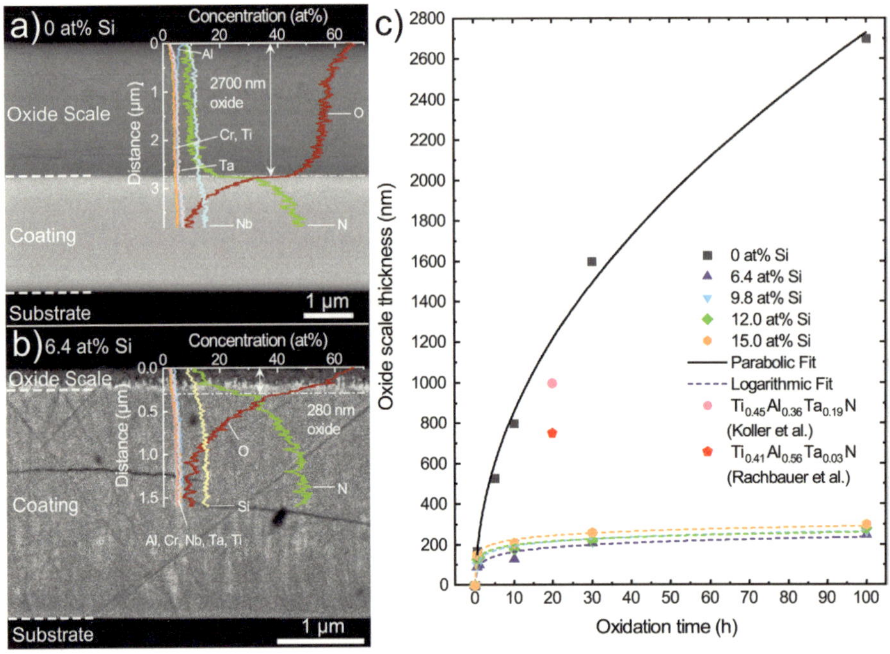

Figure 11.
SEM cross-section of the (AlCrNbTaTi)N without Si (a) and with 6.4 at.% Si (b) after exposure to ambient air at 850°C for 100 hrs. (c) The oxide scale thickness evolution as a function of time. Figure reproduced with permission from open access reference [64].

Si-free coating developed a 2700 nm thick oxide scale, whereas the Si-alloyed coatings had a much thinner oxide scale (only 280 nm). The relatively long oxidation time of 100 hours allows for accurate determination of the oxidation kinetics, derived from evaluating the time-dependent oxide scale thickness as plotted in **Figure 11c**. The oxidation kinetics for the scale growth on the Si-free sample, can be best described by parabolic behavior, while all Si-alloyed samples exhibit logarithmic growth kinetics.

A comparison of the oxidation behavior has been reported between of $(Al_{0.34}Cr_{0.22}Nb_{0.11}Si_{0.11}Ti_{0.22})_{50}N_{50}$ as HEFs and other conventional coatings (TiN, TiAlN) [66]. The HEFs were deposited at R_{N2} = 50% on a heated (415°C) substrate and biased substrate (−100 V). Regarding the oxidation behavior, the gain masse of TiN coating start to increase slowly at temperature of 500°C followed by a strong increase after 600°C. Moreover, in the case of TiAlN, the gain masse began to rise around 700°C with serious oxidation occurring after 900°C. However, $(Al_{0.34}Cr_{0.22}Nb_{0.11}Si_{0.11}Ti_{0.22})_{50}N_{50}$ showed no significant weight rise until around 1050°C, which means that its oxidation resistance is much higher that of TiN and TiAlN. This oxidation resistance improvement is due to the formation of a top layer Al_2O_3 that acts as an oxygen barrier. Besides, the Nb-containing SiO_2-rich layer was also formed, and the number of vacancies was reduced to slow the migration of oxygen atoms [66].

4. Applications

In general, high-entropy coatings have been used in many applications in different fields such as industry, automotive, biomedical, and aerospace. Below are presented some applications of HEFs, in particular those deposited by magnetron sputtering.

4.1 Cutting tools

High-entropy coatings have been used to coat the surface of cutting tools to extend the tool life and improve the integrity of the machined surface. Films with few micrometers of thickness are applied to these tools by using the PVD technique (also CVD). The performance of these cutting tools has been demonstrated in three machining processes. In fact, some high-entropy coatings can increase tool life by up to 50% in milling [67], 30% in turning [68], and 20% in drilling [69]. High-entropy coatings are relatively present as a new technology. Scientific research is in progress to develop new films with optimized properties. The first results are promising, and this kind of coating can have a significant impact on future cutting tools. Nitrides and carbides are hard films that can enhance the machining performances of the tools and reduce the friction force during the process. Chang et al. reported the study of coated tools by (AlCrNbSiTiV)N for dry turning of SCM440 steel [70].

4.2 Diffusion barrier

Peng et al. [71] investigated the architecture Cu/TiVCrZrHf on a silicon substrate that is deposited by magnetron sputtering. The results showed that the interlayered HEF slowed down the interdiffusion between Cu and Si at temperatures below 600°C. At 700°C, a failure was found because TiVCrZrHf crystallized and caused interdiffusion between the coating and copper. The coating may be a suitable barrier material for copper interconnections due to its structural and thermal resilience at temperatures below 600°C.

4.3 Other industrial uses

Khan et al. [72] investigated $AlCoCrCu_{0.5}FeNi$ deposited by RF magnetron sputtering from a stoichiometric target. The coating exhibits a mixed FCC + BCC structure, a hardness of 13 GPa, and an elastic modulus of 204 GPa. Due to their properties, the authors reported that the coating is ideal to be used in severe environments such as turbine blades in energy and aerospace applications.

Komarov et al. [73] found that the (TiHfZrVNb)N is extremely resistant to irradiation, with a hardness variation of 9 to 17% at 500 keV helium ion energy. This makes it an excellent choice for fuel cladding (fuel rods) [35, 73].

5. Conclusion

High-entropy alloys and their associated concepts have been rapidly explored and developed. Owing to their attractive physical, chemical, and mechanical features, HEA-related materials, both in bulk and coating form, have a wide range of applications. HEFs offer higher surface performance, including anti-diffusion, anti-corrosion, and oxidation resistance, compared to traditional counterparts. In this chapter, the focus is pointing out the HEFs prepared using magnetron sputtering technology and their performance in high-temperature environments. The main conclusions are drawn as follows:

The magnetron sputtering principle and the HEF preparation processing have been depicted. Standard DCMS and HiPIMS were addressed.

This chapter illustrates how various magnetron sputtering parameters impact the physicochemical properties of HEFs. The gas flow ratio (R_{N2}) and the incorporation of a third element (e.g., Si) have a considerable influence on various HEFs properties. Single and dual phases are generated.

Promising oxidation resistance HEFs can be found in alloys of the Al/W-Ti-Zr-Hf-Ta system, particularly in their nitride forms. In the one hand, the incorporation of other elements to this base alloy has a different effect on its properties: The introduction of nitrogen into the coatings decreased significantly the oxidation rate at 800°C. In the meantime, the activation energy (Ea) values followed, which observed a significant increase, reflecting the oxidation resistance improvement due to the formation of strong ionic-covalent M-N bonds. On the other hand, the incorporation of Si in FCC $(TiTaHfZrW/Al)N_x$/a-SiN_x configuration will improve the coatings' performances at high temperature. This oxidation resistance enhancement is illustrated by the E_a increase because of the formation of an amorphous SiN_x nanolayer in alternate with FCC $(AlTiZrHfTa)N_x$.

HEFs deposited by magnetron sputtering have shown promise as protective coatings for a variety of applications, including cutting tools, diffusion barriers, and some other industrial applications.

Intensive investigations are still being conducted to prepare, develop, and optimize the performance of HEFs at high temperatures. The attractive oxidation resistance properties of the coatings were reached by optimizing the various magnetron sputtering parameters. As per the aforementioned outlines, the future of HEFs has enormous promise for strengthening numerous sectors, spurring innovation, and pushing the boundaries of what is possible in material science and technology.

Acknowledgements

The authors thank the Université de Technologie de Troyes (UTT) for supporting the research.

Conflict of interest

The authors declare no conflict of interest.

Author details

Djallel Eddine Touaibia[1,2], Abdelhakim Bouissil[1,2], Sofiane Achache[1,2], Frederic Sanchette[1,2] and Mohamed El Garah[1,2]*

1 LASMIS—Laboratory of Mechanical and Materials Engineering, Nogent antenna-52, Pôle Technologique de Sud-Champagne, Nogent, France

2 Laboratoire de Recherche Correspondant (LRC) Commissariat à l'Energie Atomique et aux énergies alternatives (CEA)-LASMIS, Nogent International Centre for Coating Innovation (NICCI), Pôle Technologique de Sud-Champagne, Nogent, France

*Address all correspondence to: mohamed.el_garah@utt.fr

References

[1] Butt MA. Thin-film coating methods: A successful marriage of high-quality and cost-effectiveness—A brief exploration. Coatings. 2022;**12**:1115

[2] Krebs FC. Fabrication and processing of polymer solar cells: A review of printing and coating techniques. Solar Energy Materials and Solar Cells. 2009;**93**:394-412

[3] Weinstein SJ, Ruschak KJ. Coating flows. Annual Review of Fluid Mechanics. 2004;**36**:29-53

[4] Söderberg S, Sjöstrand M, Ljungberg B. Advances in coating technology for metal cutting tools. Metal Powder Report. 2001;**56**:24-30

[5] Rossnagel S. Thin film deposition with physical vapor deposition and related technologies. Journal of Vacuum Science & Technology A: Vacuum, Surfaces, and Films. 2003;**21**:S74-S87

[6] Leskelä M, Ritala M. Atomic layer deposition (ALD): From precursors to thin film structures. Thin Solid Films. 2002;**409**:138-146

[7] Yue TM, Xie H, Lin X, Yang H, Meng G. Microstructure of laser re-melted AlCoCrCuFeNi high entropy alloy coatings produced by plasma spraying. Entropy. 2013;**15**:2833-2845

[8] Billard A, Perry F. Pulvérisation cathodique magnétron. In: Techniques de l'ingénieur. Saint-Denis, France: Matériaux métalliques; 2005. pp. 1-17. M1654. DOI: 10.51257/a-v1-m1654

[9] Bräuer G, Szyszka B, Vergöhl M, Bandorf R. Magnetron sputtering–milestones of 30 years. Vacuum. 2010;**84**:1354-1359

[10] Cayado P, Hänisch J. High-Entropy superconducting materials. In: High Entropy Alloys - Composition and Microstructure Design [Working Title]. London, UK: IntechOpen; 16 Jul 2024. DOI: 10.5772/intechopen.1005291

[11] Li W, Liu P, Liaw PK. Microstructures and properties of high-entropy alloy films and coatings: A review. Materials Research Letters. 2018;**6**:199-229

[12] Duchaniya RK, Pandel U, Rao P. Coatings based on high entropy alloys: An overview. Materials Today: Proceedings. 2021;**44**:4467-4473

[13] Meghwal A, Anupam A, Murty B, Berndt CC, Kottada RS, Ang ASM. Thermal spray high-entropy alloy coatings: A review. Journal of Thermal Spray Technology. 2020;**29**:857-893

[14] Miracle DB, Senkov ON. A critical review of high entropy alloys and related concepts. Acta Materialia. 2017;**122**:448-511

[15] El Garah M, Sanchette F. Advanced multielement coatings, deposition, materials, applications. Coatings. 2023;**13**(1):131

[16] Yeh J-W. Alloy design strategies and future trends in high-entropy alloys. JOM. 2013;**65**:1759-1771

[17] Tsai M-H, Yeh J-W. High-entropy alloys: A critical review. Materials Research Letters. 2014;**2**:107-123

[18] Xing B, Ding Q, Jin B, Zuo X, Zhang N, Yin S. Corrosion resistance and passivation behavior of CoCrFeNi-TiAl high-entropy alloy coatings in acidic

solutions. Journal of Thermal Spray Technology. 2022;**31**:1673-1682

[19] Oses C, Toher C, Curtarolo S. High-entropy ceramics. Nature Reviews Materials. 2020;**5**:295-309

[20] Luo D, Zhou Q, Huang Z, Li Y, Liu Y, Li Q, et al. Tribological behavior of high entropy alloy coatings: A review. Coatings. 2022;**12**:1428

[21] Li W, Xie D, Li D, Zhang Y, Gao Y, Liaw PK. Mechanical behavior of high-entropy alloys. Progress in Materials Science. 2021;**118**:100777

[22] Wang X, Guo W, Fu Y. High-entropy alloys: Emerging materials for advanced functional applications. Journal of Materials Chemistry A. 2021;**9**:663-701

[23] El Garah M, Soubane D, Sanchette F. Review on mechanical and functional properties of refractory high-entropy alloy films by magnetron sputtering. Emergent Materials. 2024;**7**:77-101

[24] Touaibia DE, Achache S, Bouissil A, Ghanbaja J, Migot S, Yazdi MAP, et al. Oxidation resistance and mechanical properties of AlTiZrHfTa (-N) high entropy films deposited by reactive magnetron sputtering. Journal of Alloys and Compounds. 2023;**969**:172397

[25] Marlot A. Synthèse par pulvérisation cathodique magnétron et caractérisation de revêtement d'oxydes biocompatibles pour application aux implants dentaires en alliage de titane. France: Université de Lorraine; 2012

[26] Murty BS, Yeh J-W, Ranganathan S, Bhattacharjee PP. High-Entropy Alloys. Amsterdam, The Netherlands: Elsevier; 2019. pp. 1-12. DOI: 10.1016/B978-0-12-816067-1.00001-1. ISBN 9780128160671

[27] Chen H-C, Jan D-J, Chen C-H, Huang K-T. Bond and electrochromic properties of WO3 films deposited with horizontal DC, pulsed DC, and RF sputtering. Electrochimica Acta. 2013;**93**:307-313

[28] Fox-Rabinovich G, Wang Q, Chung Y. Adaptive hard coatings design based on the concept of self-organization during friction. In: Encyclopedia of Tribology. Boston, MA: Springer; 2013. pp. 16-23. DOI: 10.1007/978-0-387-92897-5_1180. ISBN 978-0-387-92897-5

[29] Yang S, Li X, Cooke K, Teer D. A study of TiMoN nano-multilayer coatings deposited by CFUBMSIP using DC and HIPIMS power. Applied Surface Science. 2012;**258**:2062-2067

[30] Ma Q, Li L, Xu Y, Gu J, Wang L, Xu Y. Effect of bias voltage on TiAlSiN nanocomposite coatings deposited by HiPIMS. Applied Surface Science. 2017;**392**:826-833

[31] Chang K-S, Chen K-T, Hsu C-Y, Hong P-D. Growth (AlCrNbSiTiV) N thin films on the interrupted turning and properties using DCMS and HIPIMS system. Applied Surface Science. 2018;**440**:1-7

[32] Velicu I-L, Tiron V, Porosnicu C, Burducea I, Lupu N, Stoian G, et al. Enhanced properties of tungsten thin films deposited with a novel HiPIMS approach. Applied Surface Science. 2017;**424**:397-406

[33] Helmersson U, Lattemann M, Bohlmark J, Ehiasarian AP, Gudmundsson JT. Ionized physical vapor deposition (IPVD): A review of technology and applications. Thin Solid Films. 2006;**513**:1-24

[34] Chargui A, El Beainou R, Mosset A, Euphrasie S, Potin V, Vairac P, et al.

Influence of thickness and sputtering pressure on electrical resistivity and elastic wave propagation in oriented columnar tungsten thin films. Nanomaterials. 2020;**10**:81

[35] Touaibia DE. Elaboration et caractérisation des films AlTiZrHfTa (N) déposés par pulvérisation cathodique magnétron [thesis]. Troyes: Doctoral School of Engineering Sciences; 2024

[36] Zhang Y, Yan X-H, Liao W-B, Zhao K. Effects of nitrogen content on the structure and mechanical properties of (Al0. 5CrFeNiTi0. 25) Nx high-entropy films by reactive sputtering. Entropy. 2018;**20**:624

[37] Touaibia DE, Achache S, Bouissil A, Parent F, Ghanbaja J, Gorbunova A, et al. Oxidation performance of nano-layered (AlTiZrHfTa) Nx/SiNx coatings deposited by reactive magnetron sputtering. Materials. 2024;**17**:2799

[38] Mevrel R, Pichoir R. Les revêtements par diffusion. Materials Science and Engineering. 1987;**88**:1-9

[39] Fu X, Schuh CA, Olivetti EA. Materials selection considerations for high entropy alloys. Scripta Materialia. 2017;**138**:145-150

[40] Li J, Huang Y, Meng X, Xie Y. A review on high entropy alloys coatings: Fabrication processes and property assessment. Advanced Engineering Materials. 2019;**21**:1900343

[41] Zhang T, Xin L, Wu F, Zhao R, Xiang J, Chen M, et al. Microstructure and mechanical properties of FexCoCrNiMn high-entropy alloys. Journal of Materials Science & Technology. 2019;**35**:2331-2335

[42] Owen L, Pickering E, Playford H, Stone H, Tucker M, Jones N. An assessment of the lattice strain in the CrMnFeCoNi high-entropy alloy. Acta Materialia. 2017;**122**:11-18

[43] Wang A, An X, Gu J, Wang X, Li L, Li W, et al. Effect of grain size on fatigue cracking at twin boundaries in a CoCrFeMnNi high-entropy alloy. Journal of Materials Science & Technology. 2020;**39**:1-6

[44] El Garah M, Touaibia DE, Achache S, Michau A, Sviridova E, Postnikov PS, et al. Data on nitridation effect of AlTiTaZrHf (-N) high entropy films by X-ray photoelectron spectroscopy. Data in Brief. 2022;**42**:108241

[45] Tsai D-C, Chang Z-C, Kuo B-H, Shiao M-H, Chang S-Y, Shieu F-S. Structural morphology and characterization of (AlCrMoTaTi) N coating deposited via magnetron sputtering. Applied Surface Science. 2013;**282**:789-797

[46] Yeh J-W. Overview of high-entropy alloys. In: High-Entropy Alloys: Fundamentals and Applications. Cham: Springer; 2016. pp. 1-19

[47] Braic M, Braic V, Balaceanu M, Zoita CN, Vladescu A, Grigore E. Characteristics of (TiAlCrNbY)C films deposited by reactive magnetron sputtering. Surface and Coatings Technology. 2010;**204**(12-13):2010-2014

[48] Chang Z-C, Liang S-C, Han S, Chen Y-K, Shieu F-S. Characteristics of TiVCrAlZr multi-element nitride films prepared by reactive sputtering. Nuclear Instruments and Methods in Physics Research Section B: Beam Interactions with Materials and Atoms. 2010;**268**:2504-2509

[49] Cheng K-H, Lai C-H, Lin S-J, Yeh J-W. Structural and mechanical

properties of multi-element (AlCrMoTaTiZr) Nx coatings by reactive magnetron sputtering. Thin Solid Films. 2011;**519**:3185-3190

[50] Guo Y, Shang X, Liu Q. Microstructure and properties of *in-situ* TiN reinforced laser cladding CoCr2FeNiTix high-entropy alloy composite coatings. Surface and Coatings Technology. 2018;**344**:353-358

[51] Li X, Feng Y, Liu B, Yi D, Yang X, Zhang W, et al. Influence of NbC particles on microstructure and mechanical properties of AlCoCrFeNi high-entropy alloy coatings prepared by laser cladding. Journal of Alloys and Compounds. 2019;**788**:485-494

[52] Veselkov S, Samoilova O, Shaburova N, Trofimov E. High-temperature oxidation of high-entropic alloys: A review. Materials. 2021;**14**:2595

[53] Arshad M, Amer M, Hayat Q, Janik V, Zhang X, Moradi M, et al. High-entropy coatings (HEC) for high-temperature applications: Materials, processing, and properties. Coatings. 2022;**12**:691

[54] Bouissil A, Achache S, Touaibia DE, Panicaud B, Yazdi MAP, Sanchette F, et al. Properties of a new TiTaZrHfW (N) refractory high entropy film deposited by reactive DC pulsed magnetron sputtering. Surface and Coatings Technology. 2023;**462**:129503

[55] Sheikh S, Gan L, Tsao T-K, Murakami H, Shafeie S, Guo S. Aluminizing for enhanced oxidation resistance of ductile refractory high-entropy alloys. Intermetallics. 2018;**103**:40-51

[56] Zelenitsas K, Tsakiropoulos P. Effect of Al, Cr and Ta additions on the oxidation behaviour of Nb–Ti–Si in situ composites at 800°C. Materials Science and Engineering: A. 2006;**416**:269-280

[57] Kim BG, Kim GM, Kim CJ. Oxidation behavior of TiAl-X (X = Cr, V, Si, Mo or Nb) intermetallics at elevated temperature. Scripta Metallurgica et Materialia. 1995;**33**:1117-1125

[58] del Pilar Moricca M, Varma S. High temperature oxidation characteristics of Nb–10W–XCr alloys. Journal of Alloys and Compounds. 2010;**489**:195-201

[59] Bouissil A, Achache S, Touaibia DE, Ghanbaga J, Postnikov PS, Chehimi MM, et al. Refractory high entropy TiTaZrHfW-N/Si3N4 nano-layered alloy thin film's oxidation resistance. Journal of Alloys and Compounds. 2025;**1010**:177046

[60] Yashar PC, Sproul WD. Nanometer scale multilayered hard coatings. Vacuum. 1999;**55**:179-190

[61] Chen Y-H, Lee KW, Chiou W-A, Chung Y-W, Keer LM. Synthesis and structure of smooth, superhard TiN/SiNx multilayer coatings with an equiaxed microstructure. Surface and Coatings Technology. 2001;**146**:209-214

[62] Hu X, Zhang H, Dai J, Li G, Gu M. Study on the superhardness mechanism of Ti–Si–N nanocomposite films: Influence of the thickness of the Si3N4 interfacial phase. Journal of Vacuum Science & Technology A. 2005;**23**:114-117

[63] Dong Y, Zhao W, Yue J, Li G. Crystallization of Si3N4 layers and its influences on the microstructure and mechanical properties of ZrN/Si3N4 nanomultilayers. Applied Physics Letters. 2006;**89**(12):121916

[64] Kretschmer A, Kirnbauer A, Moraes V, Primetzhofer D, Yalamanchili K,

Rudigier H, et al. Improving phase stability, hardness, and oxidation resistance of reactively magnetron sputtered (Al, Cr, Nb, Ta, Ti) N thin films by Si-alloying. Surface and Coatings Technology. 2021;**416**:127162

[65] Padamata SK, Yasinskiy A, Yanov V, Saevarsdottir G. Magnetron sputtering high-entropy alloy coatings: A mini-review. Metals. 2022;**12**:319

[66] Shen W, Tsai M, Tsai K, Juan C, Tsai C, Yeh J, et al. Superior oxidation resistance of (Al0. 34Cr0. 22Nb0. 11Si0. 11Ti0. 22) 50N50 high-entropy nitride. Journal of the Electrochemical Society. 2013;**160**:C531

[67] Constantin G, Balan E, Voiculescu I, Geanta V, Craciun V. Cutting behavior of Al0. 6CoCrFeNi high entropy alloy. Materials. 2020;**13**:4181

[68] Rizzo A, Goel S, Luisa Grilli M, Iglesias R, Jaworska L, Lapkovskis V, et al. The critical raw materials in cutting tools for machining applications: A review. Materials. 2020;**13**:1377

[69] Grigoriev SN, Migranov MS, Melnik YA, Okunkova AA, Fedorov SV, Gurin VD, et al. Application of adaptive materials and coatings to increase cutting tool performance: Efficiency in the case of composite powder high speed steel. Coatings. 2021;**11**:855

[70] Chang C-H, Yang C-B, Sung C-C, Hsu C-Y. Structure and tribological behavior of (AlCrNbSiTiV) N film deposited using direct current magnetron sputtering and high power impulse magnetron sputtering. Thin Solid Films. 2018;**668**:63-68

[71] Peng X, Chen L. Effect of high entropy alloys TiVCrZrHf barrier layer on microstructure and texture of Cu thin films. Materials Letters. 2018;**230**:5-8

[72] Khan NA, Akhavan B, Zhou H, Chang L, Wang Y, Sun L, et al. High entropy alloy thin films of AlCoCrCu0. 5FeNi with controlled microstructure. Applied Surface Science. 2019;**495**:143560

[73] Komarov F, Pogrebnyak A, Konstantinov S. Radiation resistance of high-entropy nanostructured (Ti, Hf, Zr, V, Nb) N coatings. Technical Physics. 2015;**60**:1519-1524

Chapter 2

Recent Advances on the Deposition of Thin Film Solar Cells

Siphelo Ngqoloda, Thelma Ngwenya and Mpfunzeni Raphulu

Abstract

Thin film solar cells have emerged as a promising technology in the field of photovoltaics due to their potential for reduced material usage, flexibility, and lower manufacturing costs compared to traditional crystalline silicon-based solar cells. This chapter provides an overview of thin film solar cell technology, focusing on various types such as amorphous silicon (a-Si), cadmium telluride (CdTe), copper indium gallium selenide (CIGS), and organic-inorganic perovskites-based solar cells (PSCs). The various thin film deposition techniques or methods such as vacuum-based magnetron sputtering, vacuum thermal evaporation, various chemical vapor deposition methods, and chemical solution-based methods will be extensively discussed. The focus will be on the importance of process parameters toward achieving high-quality thin films. The advantages and drawbacks of each deposition technique will be highlighted looking at their impact toward advancing thin film solar cell technology.

Keywords: thin film solar cells, vacuum thermal evaporation, sputtering, chemical vapor deposition, spin coating

1. Introduction

Over the past several decades, single or monocrystalline silicon (c-Si) solar cells have dominated the photovoltaic (PV) technology field due to their high power conversion efficiency (PCE) and better operational stability. Solar cells play an important role in the renewable energy mix to replace fossil fuels for electricity production. In brief, solar cells operate by converting solar energy directly into electrical energy. Photons (light) must reach the solar cell that has different layers such as the electron rich n-type and the electron-deficient (hole-rich) layers that are combined to form a p-n junction which is separated by a thin depletion region. Electron-hole pairs are created when light of an appropriate energy reaches the surface and is absorbed. These electron-hole pairs are separated and travel toward the respective electric contacts which are connected to an external circuit. The semiconductor layer is the most important layer in the solar cell because it absorbs photons depending on its energy band gap, generate electron-hole pairs, and transport these charge carriers toward the contacts. The operating mechanism of a typical c-Si solar cell is shown in **Figure 1** [1, 2].

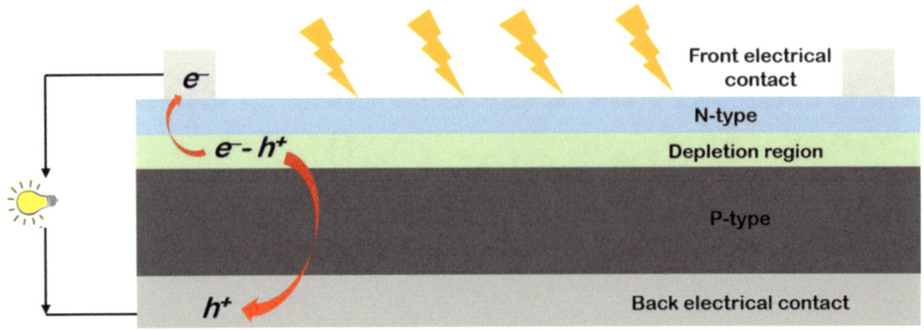

Figure 1.
Schematic diagram showing the operating mechanism of c-Si solar cell [1, 2].

To rate these solar cell devices the PCE is calculated and is defined as the fraction of solar power that a solar cell converts into electricity or simply the maximum power a solar cell can produce per given incident power. To calculate this PCE, some physical parameters are measured and used such as the short circuit current density (J_{sc}), open circuit voltage (V_{oc}), and a maximum power density leading to a fill factor (FF) a unit-less parameter. PCE is calculated from Eq. (1), using the FF that is calculated from the ratio of maximum power, that is, the product of maximum current density (J_{max}) and maximum voltage (V_{max}) based on Eq. (2):

$$PCE = \frac{J_{sc} \times V_{oc} \times FF}{P_{input}} \times 100\% \tag{1}$$

$$FF = \frac{J_{\max} \times V_{\max}}{J_{sc} \times V_{oc}} \tag{2}$$

The solar cell's PCE is normalized by its active area and the input power density. The National Renewable Energy Laboratory (NREL) in the United States of America regularly publishes the developments and breakthroughs in solar cell device efficiencies. Although the monocrystalline Si-based PV solar modules (combined solar cells) prices have been steadily declining over the years, they are still regarded as relatively expensive, especially for developing countries and regions. This is largely due to the costly process of purifying silicon from quartz such as the Czochralski process, yet, c-Si solar cells/modules still dominate the market. Crystalline Si solar cells can have wafer thicknesses of up to 350 μm which is very thick. The c-Si and the recently discovered hybrid perovskite-based solar cells have reached NREL-certified PCEs of above 26% which is the highest among other solar cell technologies as shown in **Figure 2** [3]. A single junction c-Si solar with an energy band gap of 1.1 eV is limited to 33% maximum PCE based on the Schockley-Queisser solar cell theory. This means the current PCE of 26% is close to the maximum theoretical limit. The approach to improve the PCE beyond 33% is to combine different materials that absorb different photon energies and these are known as multi-junction solar cells [1, 2].

To mitigate these cost challenges, solar cells that require less material (thinner) to absorb enough photon energies to be converted into electricity have been under intense research over the decades. These solar cell technologies are known as thin-film solar cells and are designed to convert sunlight into electricity at a lower cost. Thin-film solar cells are known as second-generation solar cells following after the first

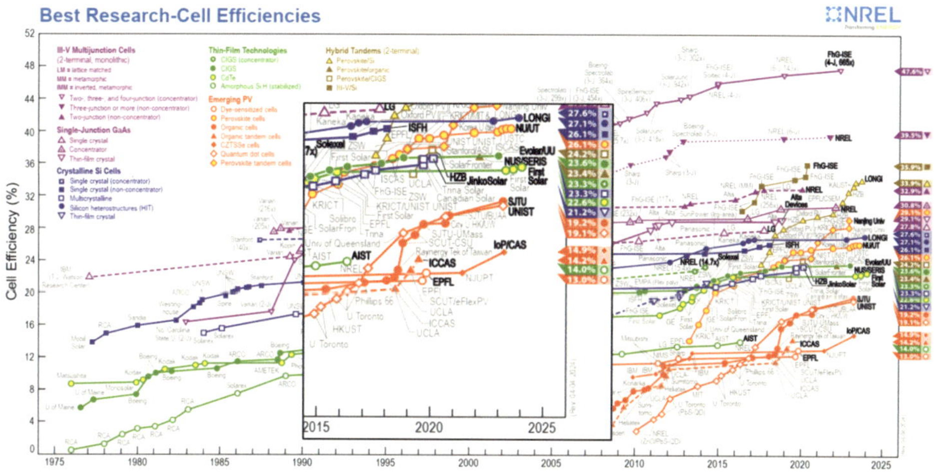

Figure 2.
National Renewable Energy Laboratory (NREL) best-certified solar cell efficiency chart [3].

generation, that is, crystalline wafer-based solar cells. There are also third-generation (emerging) solar cells that are traditionally prepared at relatively lower temperatures. Thin-film solar cells are distinguished by their thin layers of light-absorbing materials, which are much thinner than traditional monocrystalline Si solar cells. Thin film solar cells offer several advantages, such as flexibility, lightweight design, and lower material use and costs [4].

The thin semiconductor layers allow these solar cells to be more versatile, enabling applications on surfaces that would be challenging for traditional single-crystalline solar cells. However, thin film solar cells generally have lower efficiency than single crystalline solar cells such as crystalline silicon, and as a result, this technology heavily dominates the market and could continue do because of its stability compared to the emerging low-cost technologies. Due to their lightweight and flexible properties, thin film solar cells are employed in building-integrated PVs, portable devices, and vehicles. Current research on thin film solar cells is focusing on improving their efficiency and lifespan while maintaining low-cost advantages. These layers are typically only a few hundred nanometers to several micrometers thick (0.2–10 μm) and are 10 to 1000 times thinner than conventional monocrystalline Si solar cells. This reduction in material usage not only lowers costs but also allows for greater flexibility in the design and application of solar cells [5, 6].

2. Thin film deposition methods

Thin film solar cells are fabricated through layer-by-layer deposition methods that precisely deposit various thin layers required in the solar cell. Second-generation thin film solar cells are traditionally deposited through vapor-based deposition techniques such as chemical vapor deposition (CVD) and physical vapor deposition (PVD) method (**Figure 3a–c**). On the other hand, the emerging/ third-generation solar cells could be easily deposited through chemical solution methods at low temperatures (**Figure 3d**). The selection of the deposition technique is highly dependent on the chemical, physical, and thermal properties of the starting material or precursor and the desired thin film quality. For example, if the precursor is a gas such as silane

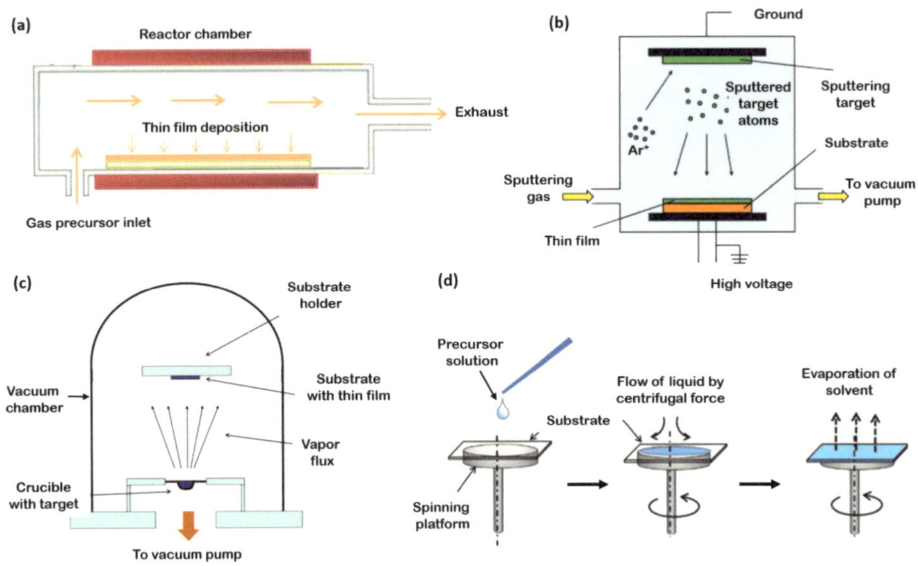

Figure 3.
Common thin film deposition techniques. (a) Chemical vapor deposition, (b) sputtering deposition, (c) thermal evaporation, (d) solution spin coating [7].

(SiH_4) which is a silicon precursor, a CVD method will be suitable. For a liquid precursor that dissociates at lower temperatures, a chemical solution method could be employed. However, for a solid material with a high melting point such as metal oxides and other inorganic compounds, a PVD method is preferred [7].

CVD is one of the mature techniques used in the deposition of thin films and coatings. It is composed of three important components, namely, the chemical vapor precursors and carrier gasses, the CVD reactor, and the effluent gas handling system (exhaust) as shown in **Figure 3a**. CVD comes in different variants, most of which are employed in thin-film solar cell depositions. These variants are thermal CVD, hot wire/filament CVD, and plasma-enhanced CVD. In thermal CVD processes, chemical reactions are activated through thermal heating using heating coils that surround the heating zone. In addition, the thermal CVD method further varies according to working pressure leading to atmospheric CVD, low-pressure CVD, and ultra-high vacuum CVD. Hot filament CVD uses thin hot filament to create reactive species on its surface. On the other hand, plasma-enhanced CVD relies on radio frequency or microwave plasma to generate reactive chemical species [8–10].

To deposit a thin film in all these CVD variants, there are certain steps that need to be achieved. These steps are the generation of gaseous reactant species, formation of intermediate reaction species, adsorption of the species by a heated substrate to initiate heterogeneous reactions at the gas-solid interface (substrate surface), formation of the film (crystallization and growth), and removal of by-products (unreacted species) to the exhaust. The type of precursor, carrier gas, gas flow rates, reactor temperature, substrate temperature, chamber pressure, and deposition time all affect the quality of the thin film [8–10].

The PVD method is different from the CVD in that it refers to the thin film deposition method where vaporized solid material's atoms or molecules condense on a solid substrate, thus forming a thin film without chemical reactions involved along the path. Some of the common methods of PVD are magnetron sputtering (direct current or radio frequency) deposition (**Figure 3b**) and vacuum thermal evaporation. During sputtering deposition, a target solid material is employed and is bombarded or sputtered with inert gas ions such as Ar ions under high vacuum conditions. Upon the sputtering of the target with ions, sputtered target molecules or atoms are directed toward a cooler substrate due to the created vacuum forming a layer-by-layer thin film. Magnetron sputtering instruments are often used for the deposition of thin films from compounds or elements with high sublimation temperatures such as metals or metal oxides [11, 12].

The frequently employed PVD method for lower-temperature compounds or metals is known as vacuum thermal evaporation and is shown in **Figure 3c**. Thermal evaporation deposition involves the heating of a powder material or in a pellet form above its sublimation or melting temperature under a high vacuum without any gas. The target material is usually contained in a boat or crucible that is heated by resistive heating. The created vacuum and lower temperature at the substrate allows easy movement of the sublimated vapor toward the substrate often happening in a straight line trajectory. The vapor condenses on the substrate surface transforming from a gaseous phase into a solid layer, which over continuous cycles forms a thin film. The versatility of this technique allows for the deposition of organic, organic-inorganic hybrid materials, and metals with low sublimation or melting temperatures from both powder and pellet targets. Although the PVD method is rather expensive due to the required high vacuum, it produces good film control in terms of thickness, uniformity, and chemical composition [8, 11, 12].

The emerging thin film solar cells are traditionally fabricated at lower temperatures at atmospheric pressures in contrast to their second-generation counterparts. This ensures low-cost fabrication and easy scalability of the deposition process. The deposition methods used are derivatives of chemical solution-based methods where precursor materials of interest are dissolved in appropriate solvents. These solution methods include spin coating (**Figure 3d**), spray coating, chemical bath, slot die coating, inkjet coating, etc. The spin coating method is more common due to its ease and low-cost instruments but difficult to scale up and therefore only suitable for small laboratory use. During the spin coating, the solution is dispensed on the surface of the substrate attached to a rotator. The solution spreads during spinning covering the entire substrate due to centrifugal force as shown in **Figure 3d**. The wet thin film is then dried to evaporate the solvent thus forming a solid thin film, which may require further heat treatment to crystallize and form the required structural phase. The formation of the film on the other chemical solution variants is similar to spin coating. In all these variants, certain aspects need to be optimized to achieve high-quality thin films such as solution viscosity, density, temperature, and solvent evaporation rate after coating [13–15].

3. Thin film solar cells

Thin film-based solar cells have evolved from the older technologies known as second-generation solar cells to recently discovered third-generation (emerging) solar cells. Second-generation thin film solar cells include hydrogenated amorphous silicon (a-Si:H) solar cells, cadmium telluride (CdTe) solar cells, and copper indium gallium selenide (CIGS) solar cells. Emerging thin film solar cells on the other hand include dye-sensitized solar cells (DSSCs), organic-based solar cells (OSCs), and organic-inorganic hybrid perovskites solar cells (PSCs) among others. While thin film solar cells require less material due to their thin construction, their PCE tends to be lower than that of conventional single crystalline-based cells. However, some second-generation thin film technologies, particularly CIGS and CdTe, have demonstrated significant efficiency improvements over the years surpassing 22% PCE, making them competitive in specific applications. Although not commercialized yet, hybrid perovskites-based solar cells have seen the fastest rise in PCE among all solar cell technologies reaching more than 26% in just over a decade of research. The high PCE of c-Si solar cells is largely attributed to the high quality (high crystalline) of the material. Similar, to achieve high PCE in thin-film solar cells, high-quality materials are required and that can be achieved through careful control of the material deposition and growth. Thin films for solar cell are usually polycrystalline and the larger the grains, the better for improved charge carrier mobility [3, 4]. As the absorber layer is more important than the other layer and forms the core of the solar cells, the subsections below will for focus more on the deposition of the absorber layer.

3.1 Amorphous silicon solar cells

Hydrogenated amorphous silicon (a-Si:H) solar cells are a type of second-generation thin film solar cell that utilizes non-crystalline (short-range order) silicon as the semiconductor material. These cells are part of a broader category of second-generation thin film photovoltaics, which also include CdTe and CIGS solar cells. Hydrogenated amorphous silicon solar cells are known for their flexibility, lightweight, and low-manufacturing costs, making them suitable for a variety of applications, particularly where traditional rigid crystalline silicon panels are not feasible. Unlike crystalline silicon, which has a well-ordered atomic structure, amorphous silicon is characterized by its disordered, or amorphous, atomic arrangement. This lack of long-range order allows for the deposition of very thin layers of silicon, typically about a micrometer thick, compared to several hundred micrometers for crystalline silicon cells. The energy band gap of a-Si:H is about 1.7 eV although there exists mid-gap states called dangling bonds causing high-defect density. Hydrogen is therefore incorporated into a-Si to cure these dangling bonds thus reducing the defects [16].

The typical structure of an a-Si:H solar cell consists of three important layers forming the so-called p-i-n junction. These layers are positively doped a-Si:H (p-type), intrinsic a-Si:H (un-doped), and negatively doped a-Si:H (n-type) layers as shown in **Figure 4a**. The role of the doped layers is to extract and transport charge carriers (electrons and holes) generated from the intrinsic layer after the absorption of a photon. These layers are deposited on a substrate that can be either rigid (e.g., glass) or flexible (e.g., plastic). Two electrodes based on the transparent conductive oxide (TCO) allow light to reach the semiconductors and a metal on the other side is employed as an electrical contact layer. TCO layers such as indium tin oxide (ITO) and fluorine-doped tin oxide (FTO) are frequently employed as well as Ag or Al metal electrodes. To maximize light absorption in a-Si:H, a back reflector is often included to reflect unabsorbed light back into the intrinsic absorber layer thus increasing the efficiency of the cell (**Figure 4a**). These different layers are deposited through different thin film deposition techniques such as plasma-enhanced CVD and hot-wire CVD for the a-Si:H layers and the back reflector, PVD-sputtering for the TCO, and thermal evaporation for the metal electrode [17, 18].

Plasma-enhanced CVD is a commonly preferred method to deposit high-quality amorphous silicon thin films. This method is preferred because the thin films are

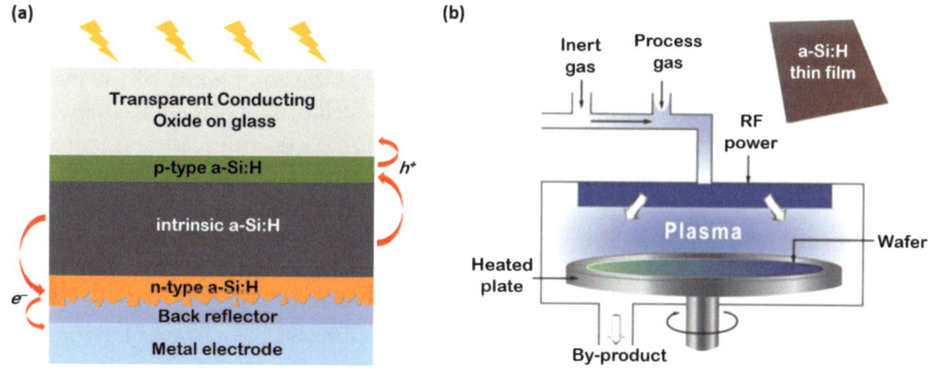

Figure 4.
(a) Structure of the p-i-n a-Si:H solar cell showing all the important thin film layers and the directions of charge transfer. (b) Schematic of a plasma-enhanced CVD system (insert is a-Si:H thin film on a substrate). Reproduced with permission from Ref. [17].

produced at low temperatures and at a short time duration. Some of the parameters that are important to consider are the Si precursor type, gas flow rates, and deposition pressure. The silicon precursor employed is silane gas (SiH_4) with H_2 used as a dopant to form a-Si:H, while argon is sometimes employed to initiate the plasma. The plasma-enhanced CVD system creates a plasma by applying a high-frequency electric field (typically in the radio frequency of 13.56 MHz) to the gas mixture (**Figure 4b**). The energy from the electric field ionizes the SiH_4 and H_2 gasses, creating a plasma composed of electrons, ions, SiH_x (x = 1–3) radicals, and other reactive species. The SiH_3 radical is particularly important as it is the primary species contributing to the growth of the a-Si film. The SiH_3 radicals, along with other reactive species, diffuse through the plasma and reach the substrate surface, where they adsorb. The substrate is typically maintained at a temperature of around 200–300°C, which is high enough to promote surface reactions of the radicals that agglomerate and coalesce to form a-Si:H thin films (**Figure 4b**, insert). In addition, to control the film thickness, the deposition rate can be controlled by adjusting the gas flow rates, plasma power, substrate temperature, and deposition pressure [19, 20].

3.2 Copper indium gallium selenide solar cells

Copper indium gallium selenide (CIGS) solar cells are a leading technology in the field of second-generation thin film photovoltaics, known for their high efficiency and versatility. CIGS solar cells are made from a semiconductor material composed of copper (Cu), indium (In), gallium (Ga), and selenium (Se) or the combination of Se and sulfur (S), which forms a thin, flexible layer capable of converting sunlight into electricity. One of the most attractive features of CIGS solar cells is their ability to achieve high conversion efficiencies, with laboratory PCEs surpassing 23%, making them competitive with traditional silicon-based solar cells; however, they are more expensive than silicon-based solar cells. The energy bandgap of CIGS can be tuned between 1.0–1.7 eV by adjusting the ratio of indium to gallium, allowing for optimization based on specific applications and environmental conditions [21].

The thin film nature of CIGS cells allows them to be deposited on various substrates, including glass, metal, and flexible materials. This flexibility opens up many applications, from traditional rooftop installations to innovative uses in building-integrated photovoltaics, portable electronics, and even wearable devices. Despite their advantages, CIGS solar cells face challenges such as the scarcity and cost of materials such as indium and gallium, as well as the complexity of the manufacturing process. However, ongoing research and development efforts are focused on improving material efficiency, reducing production costs, and scaling up manufacturing to make CIGS a more accessible and sustainable photovoltaic technology. The cost-effectiveness of the technology also stems from only a thin layer (2–2.5 μm) required to absorb enough light to achieve high PCE due to its high absorption coefficient of 10^5 cm^{-1}. Furthermore, CIGS solar cells have high radiation tolerance and are therefore suitable for space applications [22].

CIGS solar cells are composed of multilayer thin films with each layer having its specific role and the schematic structure of it is shown in **Figure 5a**. These layers are deposited on top of a soda-lime glass where different deposition methods may be used for each layer. The first layer is the molybdenum (Mo) metal electrode with a thickness of about 0.5–2 μm and is commonly deposited through the magnetron sputtering method. Due to its high reflectivity properties, this Mo layer also acts as a reflector by reflecting unabsorbed light back to the CIGS absorber layer. Furthermore, the thickness of the Mo layer is controlled to allow sodium migration from the soda-lime glass into the CIGS absorber layer to leverage the positive effect of sodium on this absorber layer. The Mo layer also allows effective intimate contact with the CIGS adjacent layer forming a $MoSe_2$ interface layer which improves ohmic contact. A cross-sectional scanning electron microscopy (SEM) image of the CIGS thin film solar cells is shown in **Figure 5b** [23, 24].

The CIGS absorber layer is then deposited using either PVD techniques or solution-based coating methods. The criteria for selecting a desired deposition method on a lab scale are often dictated by the objective of achieving better CIGS thin film control (chemical composition and thickness) and optimum device PCE. For industrial up-scaling on the other hand, besides optimum efficiency, low cost, reproducibility, and high throughput are more important. Cadmium sulfide (CdS) layer followed by intrinsic zinc oxide (i-ZnO) and Al-doped zinc oxide (Al-ZnO or AZO) layers are then deposited sequentially, and these layers are used to transport electrons from the

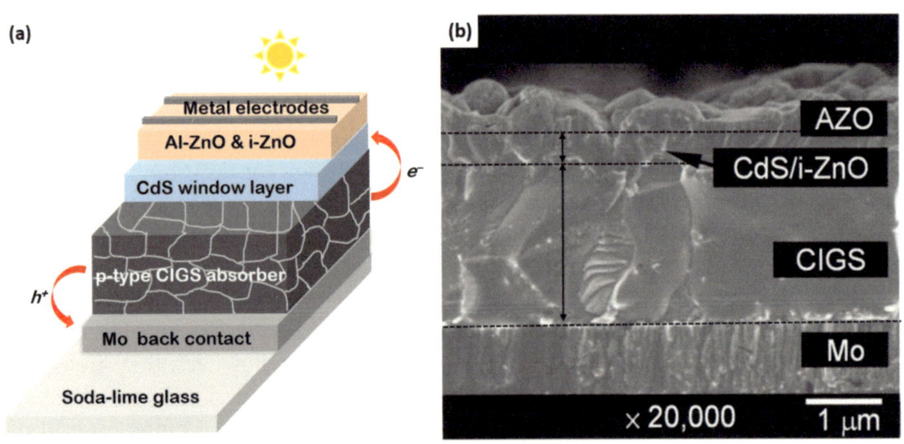

Figure 5.
(a) Structure of the CIGS solar cell showing all the important thin film layers and the directions of charge transfer. (b) Cross-sectional SEM image of a CIGS solar cell with layers deposited on soda-lime glass. Reproduced with permission from Ref. [23].

absorber layer. The CIGS solar cell is completed by metal deposition. There is no clear separation between the top layers (CdS, i-ZnO, and ZnO) in **Figure 5b** due to their thinner thicknesses, and therefore, dotted lines are used to guide the eye [25].

The PVD methods include co-evaporation of the precursor metals or sequential magnetron sputtering. The co-evaporation method involves the evaporation of the individual precursors (Cu, In, Ga, Se) at the same time either through the same or different deposition rate in order to control different aspects of the CIGS thin film. Another way is to vary the deposition rate of each precursor at different stages of the deposition. In either option though, the Se deposition rate is usually kept constant throughout. This co-evaporation method is more effective if the substrate is heated while performing the co-evaporation to ensure the film crystallizes *in situ* and Se remains in the film. X-ray fluorescence spectroscopy is commonly employed to analyze the chemical composition of CIGS thin films along the thickness before and after annealing [24, 25].

The sputtering method is known as a sequential deposition where Cu, In, and Ga are sputtered in the first step forming layers with their thicknesses optimized for best composition. The sputtered Cu/In/Ga layers are then subjected to reactive annealing in a selenium atmosphere to form the CIGS thin film in the second step. This second step is known as selenization or sulfurization when sulfur is used and Se/S powders or H_2Se/H_2S gases are employed. To fully react and crystallize the CIGS, annealing temperatures in excess of 500°C are employed. Both the co-evaporation method and the sequential deposition method have recently led to the highest CIGS-based solar cells, over 23%, with the help of silver (Ag) alloying to Cu [23, 24, 26].

Non-vacuum-based methods for CIGS offer opportunities to reduce their fabrication costs and increase their availability in the market. These non-vacuum-based methods are classified as electroless chemical bath deposition or electrodeposition and solution-based methods such as dip coating, spin coating, spray coating, or inkjet printing. During the electrodeposition, complexing agents are employed to deposit the thin film to ensure that all the reduction potentials of the precursors are decreased. Meanwhile, in the solution-based deposition methods, the precursors are dissolved in appropriate solvents before being coated on the substrate. The pre-deposited film is annealed under a Se atmosphere at elevated temperatures to fully form the CIGS thin film [27, 28].

The advantages of the non-vacuum methods over PVD methods are reduced material cost and efficient material utilization, better thin film stoichiometric control, low energy input (low temperature), better throughput, and allows use of flexible substrates (roll-to-roll deposition). However, the disadvantages of these non-vacuum methods are that they yield inhomogeneous thin films, less compact, and possess pinholes (small holes in films) that act as shunts in solar cell devices leading to poor performances. Due to these complications, CIGS solar cells based on non-vacuum deposition methods are still showing poor PCE than those fabricated from PVD-based techniques. A PCE of just over 18% is reported as the best for non-vacuum CIGS solar cells. Ag alloying has also been utilized to improve grain growth of CIGS thin film through a solution-based method achieving just over 16% PCE [27, 28].

In order to reduce trap states that often lead to a loss in open circuit voltage in CIGS solar cells a three-step co-evaporation was designed. During this process, a constant Ga composition was maintained at the upper half of the CIGS absorber layer but an increased concentration toward the Mo back contact was chosen. This process also ensured a stable energy band gap of the CIGS semiconductor thin film along its thickness. This modified three-stage deposition process was conducted at a substrate temperature of 530°C starting with poor In deposition followed by In rich, then In poor again but all the elements were deposited during these three stages of the co-evaporation process to grow 2.0–2.1-μm-thick Ag containing CIGS thin films. During this method, Ga concentration was kept constant at the upper half of the absorber layer and was strongly increased toward the back Mo layer. This changing concentration of Ga and In was done to minimize the lateral and in-depth bandgap fluctuations with the potential to reduce V_{OC} losses. The cross-sectional and the top-view SEM images demonstrate very large grains (larger than 1 μm) with perfect grain boundaries as shown in **Figure 6a, b** [26].

The sequential sputter deposition of CIGS thin films has also been reported to achieve large grains where metal precursor stacks of Cu, Ga, and In were sequentially sputtered on the Mo-coated glass substrates during the first step. The second step was the deposition of Se through thermal evaporation to selenize the stacks and thus form CIGS thin films. During this selenization step, the substrates were kept at 300°C for 15 minutes and then raised to 570°C and kept for 35 minutes to form the best thin films with large grains. A high Se evaporation rate (deposition rate) of 25 Å/s was observed to form the best CIGS thin films compared to a lower deposition rate of 15 Å/s. These results confirmed the importance of deposition parameters during thin film growth [23].

The non-vacuum deposition methods have also been explored widely for the deposition of CIGS thin films such as spin coating, screen printing, and spray coating. For the preparation of screen printing inks, the metal oxide precursors such as CuO, In_2O_3, and Ga_2O_3 are dissolved in deionized water with an addition of complexes and binders. The preferred inks should have dynamic viscosity and non-Newtonian behavior to form the best thin films. The inks are screen-printed on the substrates and then dried at 100°C for a few minutes to evaporate the solvents. The thin films are then subjected to annealing at temperatures between 400 and 500°C. The last step is selenization inside a graphite box with Se powder in a tube furnace at 550°C under the flow of 5% H_2/Ar gas to complete the formation of CIGS thin films. Produced

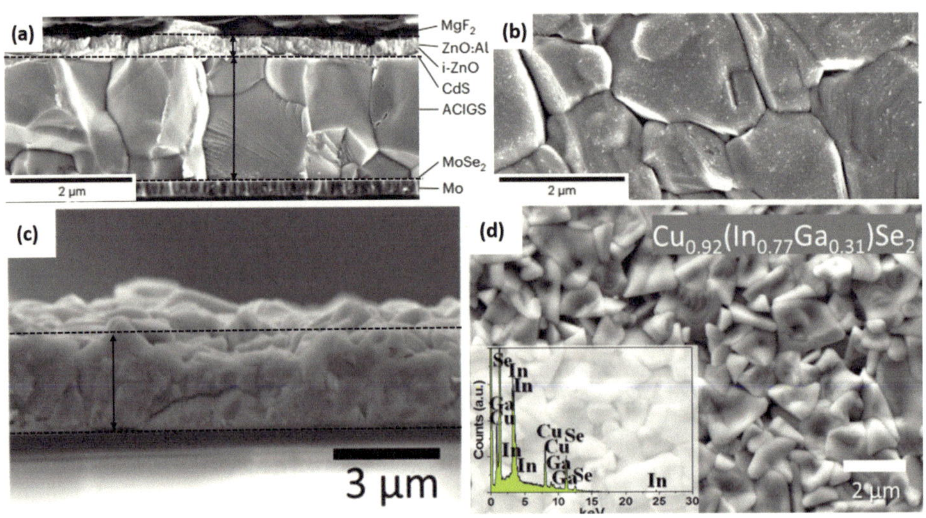

Figure 6.
(a) Cross-sectional and (b) top-view SEM images of a CIGS solar cell deposited through a co-evaporation PVD method. Reproduced with permission from ref. [26]. (c) Cross-sectional and (d) top-view SEM images of a CIGS solar cell deposited through a non-vacuum screen-printing method. Reproduced with permission from ref. [29].

thin film possess smaller grains as shown in **Figure 6c, d**. To achieve large grain *via* the non-vacuum method, metal powders (Cu, In, Ga, and Se) are used and dissolved in organic solvents of amine group and then spin-coated on substrates. To further enlarge the thin film grains, Ag was incorporated during selenization stage which infused into the thin film thus enlarging the grain sizes compared to the reference CIGS thin film. The CIGS thin films with larger grains showed better solar cell performance [28–31].

3.3 Cadmium telluride solar cells

Cadmium telluride (CdTe) solar cells significantly advance thin film-based photovoltaic technology, primarily due to their cost-effectiveness and high efficiency. As one of the leading thin film solar cell technologies, CdTe solar cells have gained a notable share of the global market, accounting for approximately 4% of the photovoltaic global market. The structure of these cells typically involves thin film layers deposited on glass or flexible substrates. These layers include a TCO layer such as FTO, cadmium sulfide (CdS) window layer, CdTe absorber layer, and lastly a metal contact. The schematic presentation of the layers is provided in **Figure 7a** and the cross-sectional SEM image of the CdTe solar cell is provided in **Figure 7b**. This configuration enables the formation of a p-n heterojunction where CdTe plays the role of a p-type material and CdS is an n-type material. This is critical for efficient charge separation and collection by the contacts (electrodes) [32, 33].

The power conversion efficiency of CdTe thin film solar cells has seen remarkable improvements over the decades, with certified PCE reaching over 22% for laboratory devices (19% in modules). This efficiency is attributed to the intrinsic properties of the CdTe semiconductor such as its direct bandgap of about 1.5 eV, which is well-suited for visible light absorption, and its high absorption coefficient, allowing for effective light capture even at minimal thicknesses. Furthermore, the lower costs associated with CdTe thin film solar cell fabrication make it an attractive alternative to traditional silicon-based solar cells, which typically require more complex and expensive fabrication processes [34, 35].

Recent advancements in CdTe thin film solar cells have focused on optimizing various parameters to enhance performance. For example, studies have shown that adjusting the absorber and buffer layer (window layer) thickness can greatly

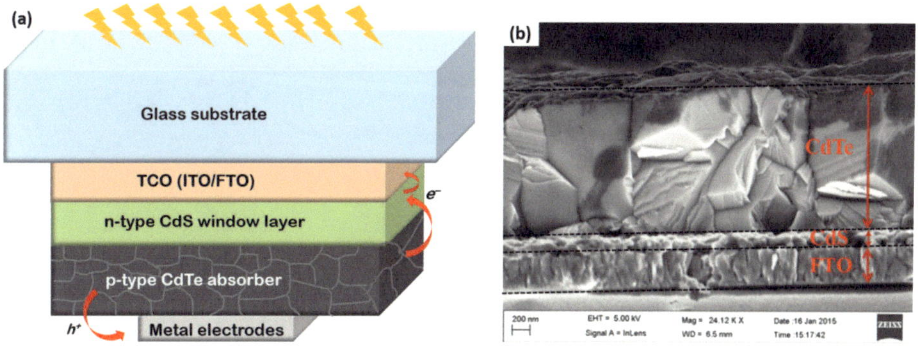

Figure 7.
(a) Structure of the CdTe solar cell showing all the important thin film layers and the directions of charge transfer. (b) Cross-sectional SEM image of a CdTe solar cell with layers deposited on FTO/glass substrate. Reproduced with permission from ref. [32].

influence the efficiency of CdTe solar cells under varying operational conditions. Innovative fabrication techniques, such as solution-based processes, have been developed to produce flexible CdTe thin film solar cells, expanding their potential applications in diverse environments. The integration of advanced materials and techniques continues to drive research in this field, aiming to overcome existing limitations and further enhance the performance and commercialization of CdTe thin film solar cells [36, 37].

The deposition methods employed for CdTe thin films significantly influence their structural, optical, and electrical properties, which in turn affect the overall performance of the solar cells. Various deposition techniques have been explored over the years, each with unique advantages and challenges. Some of the most widely employed methods for depositing CdTe thin films are vapor-based such as closed-spaced sublimation (CSS), metal-organic chemical vapor deposition (MOCVD), thermal evaporation, e-beam evaporation, sputtering, molecular beam epitaxy (MBE), and solution-based such as chemical bath deposition (CBD). The CSS method is more common because it allows for the rapid deposition of high-quality films at temperatures around 600°C, achieving thicknesses of 1–10 μm within a few minutes [38, 39].

CSS is particularly favored due to its high yield and low cost, making it a practical choice for large-scale production of CdTe thin films. The films produced by this method typically exhibit good crystallinity, which is crucial for efficient charge transport in solar cells. The CSS system is composed of a quartz tube furnace enclosed by heating lamps as shown in **Figure 8a**. To grow/deposit CdTe thin films *via* CSS, glass substrates coated with a TCO layer and CdS window layer are placed facing a CdTe powder in a graphite box (**Figure 8a**). The important deposition parameters to consider in CSS are the source and substrate temperatures, the separation distance between the source powder and the substrates, deposition pressure, and deposition time [39, 40].

High-quality CdTe thin films by CSS can be produced at a source temperature of 625°C, substrate temperature of 595°C, deposition pressure of 200 Pa, and separation distance of 2 mm. These deposition conditions produce deposition rates of 0.5–1 μm/min, meaning a deposition time of just a few minutes produces thicknesses of a few microns. The deposition mechanism of the CdTe thin films during CSS follows the fast evaporation/sublimation of CdTe powder during heating which condenses on the heated substrates sitting close. The thin film grains are quickly formed during

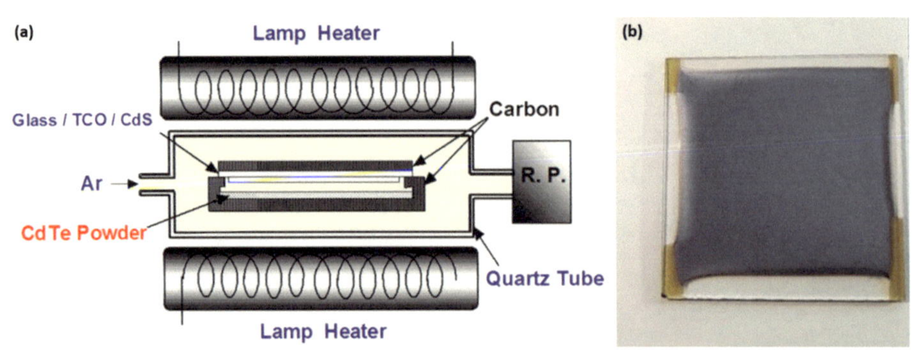

Figure 8.
(a) Schematic illustration of a closed-spaced sublimation furnace. (b) Photograph of CdTe-deposited thin film on a glass substrate. Reproduced with permission from ref. [40].

deposition leading to the formation of a thin film. Further, the deposition time can be optimized to produce the best-quality films that have large grains, compact, and pinhole-free. A picture of CdTe thin film produced *via* CSS is shown in **Figure 8b**. The deposition method has produced the best-performing CdTe solar cells with PCE of over 20% [32, 40, 41].

Although it produces high quality thin films, this high temperature, low pressure-based CSS method is still regarded expensive and therefore, low-cost deposition methods are still desirable. Low-temperature solution-based methods such as chemical bath deposition, are gaining significant interests also because they are simple and produce uniform CdTe thin films. In brief, chemical bath deposition involves the submersion of substrates to be coated in a lowly heated precursor solution containing a complex for a given time. The deposition mechanism starts with the dissociation and adsorption of the metal complex on the substrates, followed by hydrolysis of the metals and then, the evolution/growth of the solid thin films begins. The temperature and pH of the solution together with the presence of the complexing agent determines the growth mechanism of the thin film. The deposition of CdTe thin films *via* this method can be achieved at low temperatures of 65°C with a deposition time of 4 hours [42, 43].

3.4 Perovskites solar cells

Perovskite solar cells (PSCs) have emerged as a revolutionary thin film-based technology in the field of photovoltaics, characterized by their remarkably high PCEs and cost-effective fabrication processes. The rapid advancement of PSCs can be traced back to the introduction of organic-inorganic hybrid perovskites, which exhibits unique optoelectronic properties, including a tunable bandgap (1.4–2.3 eV), high absorption coefficients, and excellent charge transport capabilities. Since their inception, the efficiency of PSCs has increased from 3.8% in 2009 to certified efficiencies exceeding 26% as of 2024, demonstrating unprecedented progress in this solar cell technology. The structural properties of perovskite thin films play a crucial role in their device performance. The ideal perovskite structure, typically represented by the formula ABX_3, allows for a variety of compositions that can be engineered to optimize light absorption and charge transport. These properties can be easily tuned by carefully controlling the chosen deposition method [44].

Despite their impressive efficiencies, PSCs face significant challenges, particularly concerning structural stability and degradation when exposed to different environmental conditions. The presence of moisture and high heat have been identified as critical factors leading to the degradation of perovskite thin films, which can catalyze their decomposition. Research has focused on improving the stability of perovskite materials through various strategies, including the use of additives during crystallization, compositional engineering, and device encapsulation. Additionally, interface engineering has been explored to reduce non-radiative recombination and improve charge extraction, further enhancing the efficiency of PSCs and improving their stability. The commercial viability of PSCs is also influenced by their fabrication methods where research efforts are also looking into the development of scalable production techniques [45].

Deposition methods of perovskite thin films play a crucial role as they directly influence the morphology, crystallinity, and ultimately the performance of the solar cells. Various deposition techniques have been developed and explored over the years, each with its advantages and challenges. These deposition methods can also be classified as solution-based deposition methods and vapor-based deposition methods or their combination. Solution-based deposition methods include spin coating (one-step and two-step), spray coating, inkjet printing, slot-die, and blade coating. The vapor deposition methods on the other side include vacuum thermal evaporation (sequential and co-evaporation) and chemical vapor deposition. The hybrid method is known as the vapor-assisted solution method [46].

The composition of the perovskites absorber layer forms an ABX_3 structure where the A-site is commonly occupied by organic cation molecules such as methyl ammonium ($CH_3NH_3^+$, MA^+) or formamidinium ($CH(NH_2)_2^+$, FA^+) or inorganic cesium (Cs^+). The B-site is commonly occupied by metal cations such as lead (Pb^{2+}) or tin (Sn^{2+}) and the X-site is halides such as iodine, bromine, or a mixture of I^-, Br^-, or I^- and Cl^-. The full perovskite structure is then termed organic metal halide or all inorganic perovskites when Cs is used. Methyl ammonium lead iodide ($MAPbI_3$ or MAPI) perovskite thin films dominate in photovoltaic applications. The perovskite thin film solar cells have various structures or architectures such as n-i-p mesoporous or planar and p-i-n mesoporous or planar; the planar architectures are shown in **Figure 9a, b**. These thin film solar cells are composed of an absorber layer sandwiched by two-charge transport layers, that is, electron transport layer (ETL, n-type) and hole transport layer (HTL, p-type). The common ETLs are titanium dioxide (TiO_2) and tin dioxide (SnO_2) for the n-i-p structure and a fullerene known as PCBM in p-i-n (**Figure 9a, b**). The commonly used HTLs are Spiro-MeOTAD (large organic molecules) in n-i-p and PEDOT:PSS (polymer) or nickel oxide (NiO_x) in p-i-n devices (**Figure 9a, b**). These thin film layers that make up perovskites solar cell devices can be easily prepared at lower temperatures under atmospheric conditions [47, 48].

Perovskite thin films that are compact, pinhole-free, uniform, full surface coverage, and with larger grains are desired to ensure high efficiency and better stability. To achieve these requirements, researchers have developed different deposition techniques over the years. Solution methods are of low cost and when carefully carried they produce high-quality perovskite thin films. The simplest solution-based method is the spin coating method, which can be done either as a single step or sequential steps. In the single step (one-step), the perovskites precursors such as methyl ammonium iodide (MAI) and lead iodide (PbI_2) are dissolved in proper solvents such as dimethylformamide (DMF), dimethyl sulfoxide (DMSO), or their mixture.

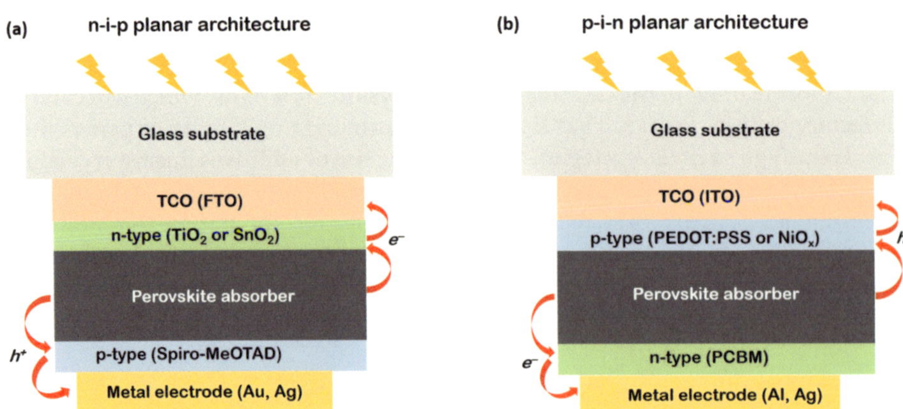

Figure 9.
Structures of perovskites solar cell, (a) conventional n-i-p planar and (b) inverted p-i-n planar.

The requirement of the solvent is to ensure the dissolution of both compounds which also has different solubility. PbI_2 is difficult to dissolve, whereas the organic halides are much easier. Poor dissolution may result in the formation of pinholes and poor surface coverage [46].

Once the solution is formed, it is dropped and spin-coated on a substrate at various speeds (3000–6000 rpm) for various time durations (20–60 seconds). During the spin coating, the solution spreads due to centrifugal force forming a wet layer of material. Spin speeds are used to adjust the thin film thickness together with solution concentration, while spin coating time may help with solvent evaporation. Thermal annealing follows the spin coating step to evaporate the solvent and crystallize the thin film. Annealing temperatures are usually low (about 100°C) and annealing time can be up to an hour, demonstrating the spin coating method is fast and facile. Slow evaporation of the solvent causes pinholes during the one-step spin coating as seen in **Figure 10a**. To mitigate this problem antisolvents are used during spin coating to quickly remove excess solvents and form crystal nuclei that help with good crystallization and the formation of a high-quality thin film. Antisolvents are non-solvents that are primarily used to precipitate solutes from a solution by reducing the solvent's power [45, 48].

The two-step deposition method on the other side involves the spin coating of the metal halide (PbI_2) dissolved in DMF or DMSO in the first step followed by spin coating of MAI in the second step. This two-step deposition method allows easier control of the perovskite thin film quality than the one-step method due to easier control of the pre-deposited PbI_2 as seen in **Figure 10b–c**. After the spin coating steps,

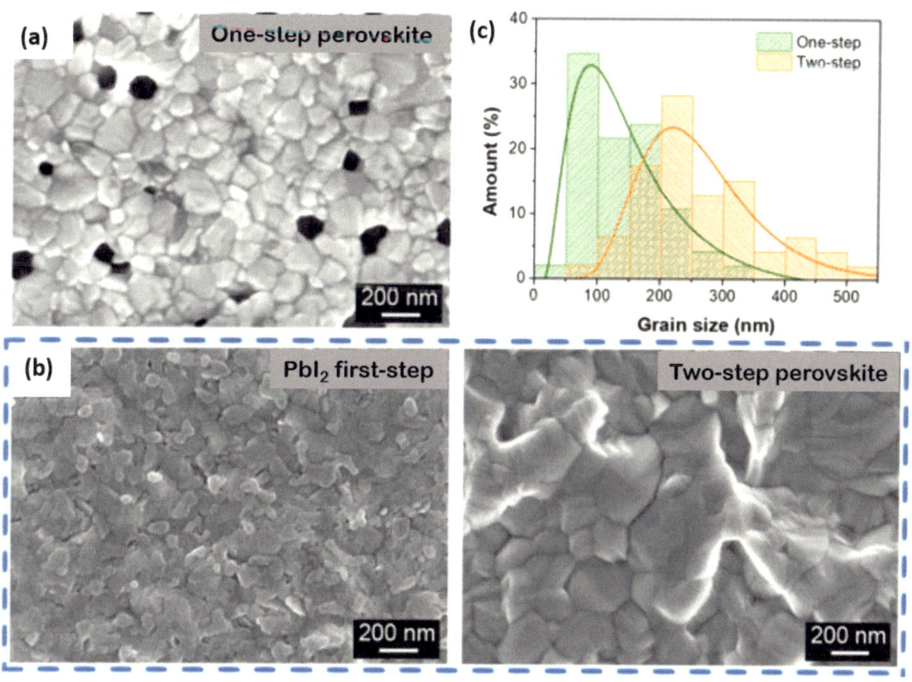

Figure 10.
(a) Top-view SEM image of a perovskite thin film by one-step method depicting large pinholes. (b) Top-view SEM images of PbI_2 (left panel) and converted perovskite (right panel) depicting compact thin films. (c) Histograms of grain sizes of perovskite thin films produced by one- and two-step methods in (a) and (b). Reproduced with permission from ref. [49].

these layers are annealed and MAI diffuses through PbI_2 and reacts to form crystalline $MAPbI_3$ perovskite thin films. One major drawback of this method is the existence of unreacted PbI_2, which can be detrimental to device efficiency by increasing charge transfer resistance at the interfaces. Strategies such as solvent annealing are commonly employed to ensure a complete reaction of PbI_2 and MAI to form high-quality perovskite thin films reaching PCE of over 25%. There are other techniques based on solution methods that are mostly scalable and are all based on dissolving perovskite precursors in solvents, coating them, and then annealing the layers to crystallize the thin films [48–50].

Vapor deposition methods such as vacuum thermal evaporation and chemical vapor deposition (CVD) offer superior control over film thickness and composition than solution-based methods. These techniques can produce uniform, high-quality perovskite films without the solvent-related issues associated with solution processing and they are easy to up-scale. The vacuum thermal evaporation deposition can be achieved through one-step co-evaporation or a sequential co-evaporation method. During the co-evaporation method, lead halides (PbI_2, $PbBr_2$, or $PbCl_2$) are evaporated at the same time as the organic halide (MAI or FAI). Although this method is fast, the complex evaporation processes of the organic halides due to their high vapor pressure raise challenges. The sequential vacuum evaporation method on the other hand offers better control where lead halides are deposited in the first step followed by evaporation of organic halides that react with pre-deposited metal halides and crystallize to form high-quality perovskite thin films. The post-annealing step is often not required during the vacuum evaporation methods. The disadvantage of this method is the requirement for a high vacuum necessary to extend the mean free path of the evaporated molecules. Another one is low yield at the substrates due to vapor condensing on the walls of the vacuum chambers [51, 52].

The CVD method does not require a high vacuum compared to thermal evaporation and the deposition of perovskite thin films is achievable by the two-step hybrid, two-step CVD, or one-step CVD methods. The two-step hybrid method involves the deposition of lead halide by either spin coating or thermal evaporation in the first step. To form a perovskite phase, these pre-deposited lead halides are exposed to a vapor of organic halide inside a tube furnace with the help of a carrier gas. The two-step CVD method involves the deposition of lead halides in a CVD tube furnace and then exposing the lead halide thin films to an organic halide vapor. On the other hand, during the one-step deposition, both lead halides and organic halides are co-evaporated inside the tube furnace and deposited on a substrate sitting downstream [52–54].

The important parameters to consider in CVD deposition are the source and substrate temperatures, source-to-substrate distance, deposition pressure, and deposition time. The substrates are usually positioned downstream in a tube sitting at a lower temperature than the source to provide a temperature gradient. To form perovskite thin films, the organic halide molecules diffuse through the lead halide structure, and the intercalation reaction begins until no lead halide phase remains. The CVD method often produces high-quality perovskite thin films with very large grains, which is beneficial for high-efficiency solar cells [52–54].

4. Conclusions

This book chapter discussed the importance of deposition methods and techniques for thin film solar cells. The focus was on the deposition of the thin film absorber layers of the solar cells as this is regarded as the important layer. The most employed deposition techniques in thin film solar cells are chemical vapor deposition, magnetron sputtering, vacuum thermal evaporation, electroless deposition, and solution-based deposition. The vapor-based techniques usually produce high-quality thin films to the nature of deposition which is layer-by-layer deposition of molecules or atoms. The major challenges associated with these vapor-based techniques are the requirements of low-pressure (vacuum) and higher temperatures leading to higher costs. Chemical solution-based techniques are often conducted at low temperatures resulting in low-energy requirements meaning low costs. The challenge, however, is the poor thin film quality that often results from solution-based deposition methods. More research and development are still necessary to improve these solution-based methods in terms of producing high-quality thin films and up-scalable.

Acknowledgements

The authors would like to thank the support from the Catalysis Group of the Advanced Materials Division at Mintek. The authors also acknowledge the financial support received from Mintek Science Vote grant.

Conflict of interest

The authors declare no conflict of interest.

Author details

Siphelo Ngqoloda*, Thelma Ngwenya and Mpfunzeni Raphulu
Advanced Materials Division, Mintek, Randburg, Gauteng Province, South Africa

*Address all correspondence to: siphelon@mintek.co.za

References

[1] Benda V, Černá L. PV cells and modules – State of the art, limits and trends. Heliyon. 2020;**6**:e05666. DOI: 10.1016/j.heliyon.2020.e05666

[2] Pastuszak J, Węgierek P. Photovoltaic cell generations and current research directions for their development. Materials. 2022;**1**:5542. DOI: 10.3390/ma15165542

[3] NREL. Available from: https://www.nrel.gov/pv/assets/pdfs/best-research-cell-efficiencies.pdf

[4] Maalouf A, Okoroafor T, Jehl Z, et al. A comprehensive review on life cycle assessment of commercial and emerging thin film solar cell systems. Renewable and Sustainable Energy Reviews. 2023;**186**:113652. DOI: 10.1016/j.rser.2023.113652

[5] Ramanujam J, Bishop DM, Todorov TK, et al. Flexible CIGS, CdTe and a-Si:H based thin film solar cells: A review. Progress in Materials Science. 2020;**110**:100619. DOI: 10.1016/j.pmatsci.2019.100619

[6] Rehman F, Syed IH, Khanam S, et al. Fourth-generation solar cells: A review. Energy Advances. 2023;**2**:1239-1262

[7] Kowsar A, Rahaman M, Saidul Islam M, et al. Progress in major thin film solar cells: Growth technologies. Layer materials and efficiencies. International Journal of Renewable Energy Research. 2019;**9**:579-597

[8] Abegunde OO, Akinlabi ET, Oladijo OP, et al. Overview of thin film deposition techniques. AIMS Materials Science. 2019;**6**:174-199

[9] Sun L, Yuan G, Gao L, et al. Chemical vapour deposition. Nature Reviews Methods Primers. 2021;**1**:5. DOI: 10.1038/s43586-020-00005-y

[10] Choy K. Chemical vapour deposition of coatings. Progress in Materials Science. 2003;**48**:57-170. Available from: www.elsevier.com/locate/pmatsci

[11] Ichou H, Arrousse N, Berdimurodov E, et al. Exploring the advancements in physical vapor deposition coating: A review. Journal of Bio- and Tribo-Corrosion. 2024;**10**:3. DOI: 10.1007/s40735-023-00806-0

[12] Ghazal H, Sohail N. Sputtering Deposition. Available from: www.intechopen.com

[13] Hossain MI, Mansour S. A critical overview of thin films coating technologies for energy applications. Cogent Engineering. 2023;**10**:2179467. DOI: 10.1080/23311916.2023.2179467

[14] Chaudhary KT. Thin Film Deposition: Solution Based Approach. Available from: www.intechopen.com (200AD)

[15] Yunus Y, Mahadzir NA, Ansari MNM, et al. Review of the common deposition methods of thin film Pentacene, its derivatives, and their performance. Polymers. 2022;**14**:1112. DOI: 10.3390/polym14061112

[16] Stuckelberger M, Biron R, Wyrsch N, et al. Review: Progress in solar cells from hydrogenated amorphous silicon. Renewable and Sustainable Energy Reviews. 2017;**76**:1497-1523

[17] Prayogi S, Ayunis A, Cahyono Y, et al. N-type H_2-doped amorphous silicon layer for solar-cell application. Mater Renew Sustain Energy. 2023;**12**:95-104

[18] Sreejith S, Ajayan J, Kollem S, et al. A comprehensive review on thin film amorphous silicon solar cells. SILICON. 2022;**14**:8277-8293

[19] Marvi Z, Xu S, Foroutan G, et al. Plasma-deposited hydrogenated amorphous silicon films: Multiscale modelling reveals key processes. RSC Advances. 2017;7:19189-19196

[20] Pandey A, Bhattacharya S, Panigrahi J, et al. Investigation of dual intrinsic a-Si:H films for crystalline silicon surface passivation by spectroscopic ellipsometry: Application in silicon heterojunction solar cells. Applied Physics A: Materials Science & Processing. 2023;**129**:575. DOI: 10.1007/s00339-023-06854-0

[21] Nwakanma O, Velumani S, Morales-Acevedo A. Review on the effects due to alkali metals on copper–indium–gallium–selenide solar cells. Materials Today Energy. 2021;**20**:100617. DOI: 10.1016/j.mtener.2020.100617

[22] Ramanujam J, Singh UP. Copper indium gallium selenide based solar cells - a review. Energy and Environmental Science. 2017;**10**:1306-1319

[23] Huang CH, Chuang WJ, Lin CP, et al. Deposition technologies of high-efficiency cigs solar cells: Development of two-step and co-evaporation processes. Crystals. 2018;**8**:296. DOI: 10.3390/cryst8070296

[24] Machkih K, Oubaki R, Makha M. A review of CIGS thin film semiconductor deposition via sputtering and thermal evaporation for solar cell applications. Coatings. 2024;**14**:1088

[25] Zhao C, Yu S, Tang W, et al. Advances in CIGS thin film solar cells with emphasis on the alkali element post-deposition treatment. Materials Reports: Energy. 2023;**3**:100214. DOI: 10.1016/j.matre.2023.100214

[26] Keller J, Kiselman K, Donzel-Gargand O, et al. High-concentration silver alloying and steep back-contact gallium grading enabling copper indium gallium selenide solar cell with 23.6% efficiency. Nature Energy. 2024;**9**:467-478

[27] Mufti N, Amrillah T, Taufiq A, et al. Review of CIGS-based solar cells manufacturing by structural engineering. Solar Energy. 2020;**207**:1146-1157

[28] Xu M, Yan S, Liang T, et al. 16.48% efficient solution-processed CIGS solar cells with crystal growth and defects engineering enabled by Ag doping strategy. Journal of Energy Chemistry. 2025;**100**:59-65

[29] Gonçalves BF, Sousa V, Virtuoso J, et al. Towards all-non-vacuum-processed photovoltaic systems: A water-based screen-printed Cu(In,Ga)Se_2 Photoabsorber with a 6.6% efficiency. Nanomaterials. 2023;**13**:1920. DOI: 10.3390/nano13131920

[30] Gonçalves BF, Botelho G, Lanceros-Méndez S, et al. Eco-friendly and cost-efficient inks for screen-printed fabrication of copper indium gallium diselenide photoabsorber thin films. Journal of Colloid and Interface Science. 2021;**598**:388-397

[31] Sousa V, Gonçalves BF, Rosen YS, et al. Over 6% efficient Cu(In,Ga)Se_2 solar cell screen-printed from oxides on FTO. ACS Applied Energy Materials. 2020;**3**:3120-3126

[32] Liang J, Lin Q, Li H, et al. Rectification and tunneling effects enabled by Al_2O_3 atomic layer deposited on back contact of CdTe solar cells. Applied Physics Letters. 2015;**107**:013907. DOI: 10.1063/1.4926601

[33] Scarpulla MA, McCandless B, Phillips AB, et al. CdTe-based thin film photovoltaics: Recent advances, current challenges and future prospects. Solar Energy Materials and Solar Cells. 2023;**255**:112289. DOI: 10.1016/j.solmat.2023.112289

[34] Ahmad NI, Kar YB, Doroody C, et al. A comprehensive review of flexible cadmium telluride solar cells with back surface field layer. Heliyon. 2023;**9**:e21622. DOI: 10.1016/j.heliyon.2023.e21622

[35] Dharmadasa IM, Alam AE, Ojo AA, et al. Scientific complications and controversies noted in the field of CdS/CdTe thin film solar cells and the way forward for further development. Journal of Materials Science: Materials in Electronics. 2019;**30**:20330-20344

[36] Harif MN, Rahman KS, Rosly HN, et al. An approach to alternative post-deposition treatment in CdTe thin films for solar cell application. Superlattices and Microstructures. 2020;**147**:106687. DOI: 10.1016/j.spmi.2020.106687

[37] Kumarasinghe RKKGRG, Kumarasinghe PKK, Wijesundera RP, et al. Thermally evaporated CdS/CdTe thin film solar cells: Optimization of $CdCl_2$ evaporation treatment on absorber layer. Current Applied Physics. 2022;**33**:33-40

[38] Sinha T, Lilhare D, Khare A. A review on the improvement in performance of CdTe/CdS thin film solar cells through optimization of structural parameters. Journal of Materials Science. 2019;**54**:12189-12205

[39] Amin N, Karim MR, Alothman ZA. An In-depth analysis of CdTe thin film deposition on ultra-thin glass substrates via close-spaced sublimation (CSS). Coatings. 2022;**12**:589. DOI: 10.3390/coatings12050589

[40] Rahman KS, Harif MN, Rosly HN, et al. Influence of deposition time in CdTe thin film properties grown by close-spaced sublimation (CSS) for photovoltaic application. Results in Physics. 2019;**14**:102371. DOI: 10.1016/j.rinp.2019.102371

[41] Li DB, Bista SS, Awni RA, et al. 20%-efficient polycrystalline Cd(Se,Te) thin film solar cells with compositional gradient near the front junction. Nature Communications. 2022;**13**:7849. DOI: 10.1038/s41467-022-35442-8

[42] Sengupta S, Aggarwal R, Raula M. A review on chemical bath deposition of metal chalcogenide thin films for heterojunction solar cells. Journal of Materials Research. 2023;**38**:142-153

[43] Kachhia ZS, Chaki SH, Patel SR, et al. Chemical bath deposited CdTe thin film: Optical, electrical, and photoresponse aspects. Next Materials. 2024;**3**:100152

[44] Basumatary P, Agarwal P. A short review on progress in perovskite solar cells. Materials Research Bulletin. 2022;**149**:111700. DOI: 10.1016/j.materresbull.2021.111700

[45] Chen Y, Zhang M, Li F, et al. Recent Progress in perovskite solar cells: Status and future. Coatings. 2023;**13**:644. DOI: 10.3390/coatings13030644

[46] Swartwout R, Hoerantner MT, Bulović V. Scalable deposition methods for large-area production of perovskite thin films. Energy and Environmental Materials. 2019;**2**:119-145

[47] Sharif R, Khalid A, Ahmad SW, et al. A comprehensive review of the current progresses and material advances in perovskite solar cells. Nanoscale Advances. 2023;**5**:3803-3833

[48] Liu S, Biju VP, Qi Y, et al. Recent progress in the development of high-efficiency inverted perovskite solar cells. NPG Asia Materials. 2023;**15**:27. DOI: 10.1038/s41427-023-00474-z

[49] Mohamad, Noh MF, Arzaee NA, Nawas Mumthas IN, et al. Superiority of two-step deposition over one-step deposition for perovskite solar cells processed in high humidity atmosphere. Optical Materials. 2021;**118**:111288. DOI: 10.1016/j.optmat.2021.111288

[50] Khorasani A, Mohamadkhani F, Marandi M, et al. Opportunities, challenges, and strategies for scalable deposition of metal halide perovskite solar cells and modules. Advanced Energy and Sustainability Research. 2024;**5**:2300275. DOI: 10.1002/aesr.202300275

[51] Yadav C, Kumar S. Review on perovskite solar cells via vacuum and non-vacuum solution based methods. Results in Surfaces and Interfaces. 2024;**14**:100210. DOI: 10.1016/j.rsurfi.2024.100210

[52] Abzieher T, Moore DT, Roß M, et al. Vapor phase deposition of perovskite photovoltaics: Short track to commercialization? Energy and Environmental Science. 2024;**17**:1645-1663

[53] Jiang Y, He S, Qiu L, et al. Perovskite solar cells by vapor deposition based and assisted methods. Applied Physics Reviews. 2022;**9**:021305. DOI: 10.1063/5.0085221

[54] Hwang JK, Jeong SH, Kim D, et al. A review on dry deposition techniques: Pathways to enhanced perovskite solar cells. Energies. 2023;**16**:5977. DOI: 10.3390/en16165977

Chapter 3

Spray Pyrolysis Technique for Cu_2MgSnS_4 Thin Films

Amal Hammoud, Badriyah Alhalaili, Ruxandra Vidu and Najoua Kamoun-Turki

Abstract

This chapter deals with the synthesis of quaternary Cu_2MgSnS_4 (CMTS) thin films, through the Spray Pyrolysis technique (SP). SP is a simple, cost-effective deposition technique that depends upon various factors. Furthermore, the advantages and disadvantages of the SP method have been explored. Moreover, the discussion has been made through the case study of the influence of deposition parameters, particularly substrate temperature, on the structural, morphological, optical, and electrical properties of CMTS thin films. The crystallization of CMTS in the kesterite phase at 300°C leads to the improvement in crystallinity and grain morphology, and favorable optical properties such as a high absorption coefficient and a suitable band gap energy of 1.47 eV. Lastly, the application of doping by Ag using Spary Pyrolysis on the physical properties of Cu_2MgSnS_4 thin films is briefly discussed, highlighting their potential for antibacterial activity. This chapter can open a new avenue for researchers working in the field by combining a general understanding of CMTS thin films with particular experimental insights.

Keywords: Cu_2MgSnS_4 (CMTS), material characterization, spray pyrolysis, antibacterial activity, thin film

1. Introduction

Cu_2MgSnS_4 (CMTS) is a semiconductor from the chalcogenide family, originating from zinc-blende or wurtzite binary II–VI chalcogenide [1]. It has various structures based on cation ordering, including stannite, kesterite, PMCA, wurtzite stannite, and wurtzite kesterite. The unit cell in CMTS contains four Cu, two Mg, two Sn, and eight S atoms. The most stable CMTS is found in the stannite phase, followed by Kesterite and PMCA. Experimentally, nanocrystals of Cu_2MgSnS_4 were synthesized in 2014 by Wei et al. [2] with kesterite phase formation found in various Zn/Mg ratios. Guo [3] prepared a tetragonal CMTS thin film, while Sharma et al. [4] showed the effect of temperature variation on CMTS thin films' structural properties. Ali et al. [5] synthesized CMTS nanoparticles with different molarities of magnesium, resulting in a cubic system with space group Fd-3 m. Yang et al. [6] suggested that CMTS crystallized in a single zinc-blende structure.

 IntechOpen

2. Application of CMTS compound

CMTS compound can be a good material for several applications, as presented in schematic **Figure 1**, such as energy conversion (photocatalysis [5], photovoltaic [6], water splitting [4]), humidity sensing [7], and biomedical applications (antibacterial activity) [8, 9]. It has gained significant attention due to its unique features, which include:

- *High absorption coefficient*: a high absorption coefficient exceeding 10^4 Cm^{-1}, which ensures efficient light absorption [10].
- *Tunable bandgap energy*: CMTS exhibits an ideal optical bandgap in the range of 1.3–1.5 eV, making it suitable for solar energy conversion [2, 4, 5, 11].
- *Low toxicity and earth abundance*: based on abundant and nontoxic elements, which eliminate the use of rare or toxic elements like indium and cadmium CdTe or CIGS [7].

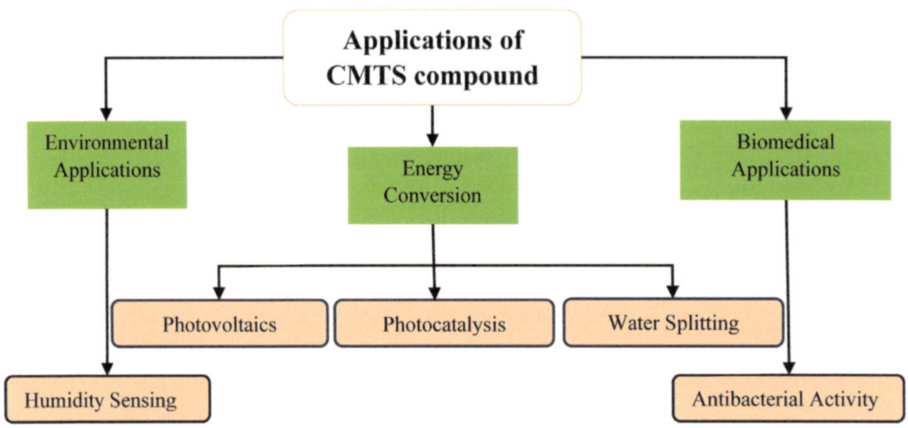

Figure 1.
Schematic presentation of CMTS compound applications.

2.1 Photocatalysis

Photocatalysis is a phenomenon including reactions that take place using light and a semiconductor in which an electron-hole pair is generated [12]. These pairs require a reduction and oxidation reaction as depicted in a series of eqs. (1)–(11) [13]. This process yields superoxide anions and hydroxyl radicals. Equations from (1) to (5) present the pollution degradation. Then, eqs. (6) and (7) refer to water splitting process and from (8) to (11) define the conversion of CO_2 to CH_4.

Pollutant degradation:

$$Sc + h\nu \rightarrow e^- + h^+ \tag{1}$$

$$e^- + O_2 \rightarrow O_2 \tag{2}$$

$$h^- + OH- \rightarrow {}^{\cdot}OH \tag{3}$$

$$O_2 + \text{Pollutant} \rightarrow \text{deg.product} \tag{4}$$

$$OH + \text{Pollutant} \rightarrow \text{deg.product} \tag{5}$$

Water splitting:

$$4e^- + 4H^+ \rightarrow 2H_2 \tag{6}$$

$$4h^+ + 2H_2O \rightarrow O_2 + 4H^+ \tag{7}$$

CO_2 conversion into CH_4:

$$CO_2 + 2H^+ + 2e^- \rightarrow HCOOH \tag{8}$$

$$HCOOH + 2H^+ + 2e^- \rightarrow HCHO + H_2O \tag{9}$$

$$HCHO + 2H^+ + 2e^- \rightarrow CH_3OH \tag{10}$$

$$CH_3OH + 2H^+ + 2e^- \rightarrow CH_4 + H_2O \tag{11}$$

2.2 Photovoltaics

Photovoltaics (PV) converts light into electricity using semiconducting materials with the photovoltaic effect [14]. The best conversion efficiency is about 0.78% for films with sulfurization temperatures of 530°C [6]. The film shows good crystalline quality and a continuous surface, which is beneficial in charge carrier transport.

2.3 Water splitting

Water splitting is a separation process of water into its elements, hydrogen and oxygen, using power sources (electrical, thermal, and light) [15]. Electrolytic water splitting is driven by passing the electrical current through the water, where conversion of the electrical energy to chemical energy takes place at the electrode solution interface through charge transfer reactions. Water reacts at the anode to form oxygen and protons, whereas a hydrogen evolution reaction takes place at the cathode [4].

2.4 Antibacterial activities

Antibacterial activity is associated with compounds or materials that can kill bacteria or slow their growth [16]. Gram-positive bacteria have a thick mesh-like peptidoglycan cell wall, while Gram-negative bacteria have a thinner one. AgNPs have been used in various antibacterial applications, with mechanisms based on smaller nanoparticles having more antibacterial activity due to large surface exposure to the bacterial membrane, the interaction of Ag + with the negative charge on the cell wall, and Ag^{+2} inhibiting protein structures by binding to thiol and amino groups.

3. Characterizations of CMTS compound

To date, a number of advanced techniques have been employed in the characterization of CMTS thin films, all of which have contributed greatly to understanding their structural, morphological, optical, and compositional properties. X-ray diffraction (XRD) has been one of the major techniques for establishing the crystalline structure and phase purity of CMTS thin films, where diffraction patterns reveal the atomic arrangement along crystallographic planes [17]. The Raman technique was performed in addition to the XRD since it provided valuable vibrational data related to specific chemical bonding and phase identity [18].

All these techniques will be adopted to characterize surface morphology, grain size, and the cross-section structure of films: Scanning electron microscopy (SEM) [19] and transmission electron microscopy (TEM) [20]. Particularly, TEM permits the examination of atomic orderings or nanostructures inside the film at high resolution. Optical properties, such as bandgap energy, are normally investigated by UV-visible spectroscopy [21], which detects light absorption in a wide spectral range and finds the electronic transitions in the material. All these techniques together provide knowledge of the properties of CMTS thin films and are extremely important in optimizing their performances in solar cells and many other applications (**Figure 2**).

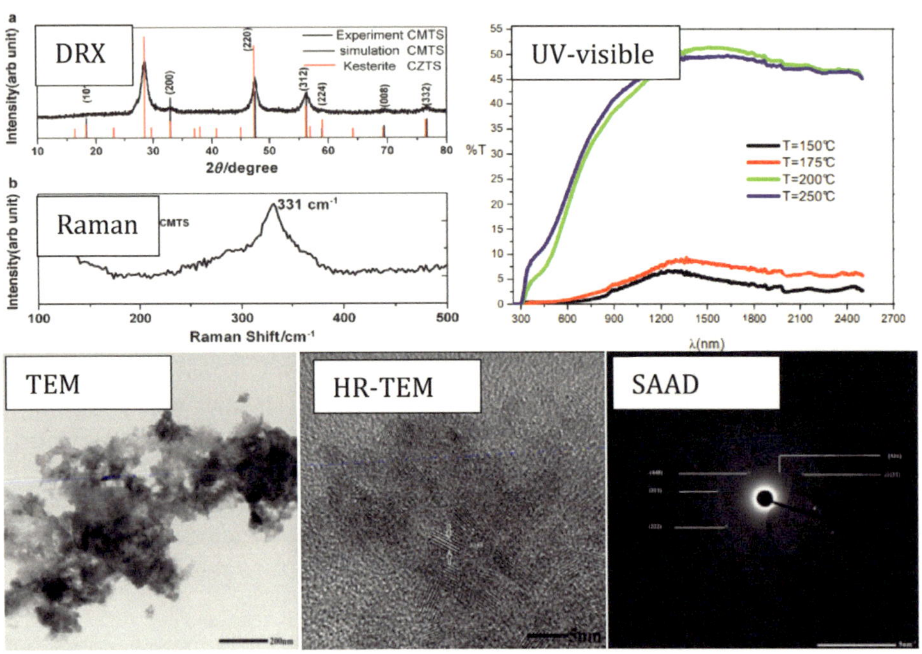

Figure 2.
Characterization used for CMTS compound.

4. Synthesis technique: Spray pyrolysis

Thin films are a crucial component of modern material science and technology, finding applications in a wide range of fields, including photovoltaics [22], optoelectronics [23], and energy storage [24]. These films are typically deposited in thicknesses ranging from a few nanometers to several micrometers and are known for their unique physical, optical, and chemical properties that differ significantly from bulk materials. The properties of thin films are determined not only by their composition but also by the deposition techniques used. In the literature, efforts to develop thin films using various techniques, employed to synthesize different types of thin film materials, have been widely documented. There are two main deposition methods commonly used for thin film fabrication: physical and chemical methods [25]. (i) Physical Vapor Deposition (PVD) involves the deposition of the material vapor in a low-pressure environment, with popular methods including sputtering [26], thermal evaporation, and pulsed laser deposition (PLD) [27]. (ii) Chemical Vapor Deposition (CVD) relies on the chemical reaction of a precursor with a substrate. Well-known CVD techniques include sol-gel [28], dip coating [29], spin coating [30], and spray pyrolysis (SP) [10, 31, 32], which is the focus of this work.

Spray Pyrolysis is a cost-effective thin film deposition method, particularly advantageous compared to other deposition techniques. It involves spraying a solution containing precursors onto heated substrates [33], where pyrolytic reactions occur due to the high substrate temperature. As a result, the solvent evaporates, and salts are strongly adsorbed onto the substrate. The experimental device of this technique is composed of:

4.1 Heating plate

The heating plate used is a block of parallel pedal shape of cast iron to ensure as perfect contact as possible with the substrates. The plate is heated by resistors placed in parallel with a power of 500 W and of the Vulcanic brand. A temperature regulation loop of the substrate Ts composed of a thermocouple housed on the heating plate and a stabilized power supply makes it possible to regulate the heating temperature Tr by varying the current transferred.

4.2 Suction pump

The “Masterflexcosole drive” type suction pump controls the flow rate of the spray solution.

4.3 Jet

A nozzle consisting of a line for liquids and another for gas allows the prepared solution to be projected onto the heated substrates. This nozzle is mounted in a nozzle that sweeps the substrates according to the (x, y) plane. The choice of gas flow is conditioned by the distance between the substrates and the nozzle. This distance affects the size of the droplets that arrive on the surface of the substrates.

4.4 Deposition steps

This process can be divided into three main steps:

4.4.1 Preparation of the spraying solution

The starting solution is formed by precursors dissolved in the solvent with a well-determined stoichiometric ratio. Distilled water or alcohol (methanol, ethanol, etc.) is often used as a solvent to obtain homogeneous and transparent solutions. Gunavathy et al. have studied the effect of solvent on the physical properties of Cu_2ZnSnS_4 absorber thin films prepared by spray pyrolysis [34]. In the case study, we use a total volume of 300 mL of bi-distilled water. However, in our previous works, we used ethanolic solvent [7]. The selected chemical precursors used as the source for sulfur are presented in **Table 1**.

4.4.2 Transport of the resultant droplets

A conical-shaped jet is generated by the nozzle where the droplets are projected onto the substrate. The size of the spray droplets and the concentration of the solution are important parameters for the homogeneity of the deposited material. Scanning speed, solution flow rate, and compressed air are also important factors in obtaining good morphology and properly transmitting the starting solution to the substrate. Ajili et al. [35] have reported the effect of flow rate on the growth of SnO_2:F thin films synthesized via Spray Pyrolysis.

4.4.3 Pyrolysis of the droplets

The formation of the films is done by evaporation of the liquid phase of the droplets on the surface of the heated substrate, followed by an endothermic reaction that produces the solid material on the substrate. Thanks to the high temperature of the substrate, the vapors formed around the droplets are released, allowing the material to bond effectively to the surface. Hammoud et al. [7] have reported the impact of substrate temperature on sprayed CMTS thin films. Here, we varied the substrate temperature (Ts) using a temperature controller, setting it to 210, 240, 270, and 300°C (**Figure 3**).

4.5 Advantages and disadvantages of spray pyrolysis

4.5.1 Advantages

- Spray pyrolysis is a low-cost thin film deposition technique that can produce different types of surfaces of films, such as uniform, porous, well-adherent,

Formula	Name	Concentration (mol/L)
$CuCl_2.2H_2O$	Cupric chloride	[Cu] =10.10^{-3}
$MgCl_2.2H_2O$	Magnesium chloride	[Mg] = 5.10^{-3}
$SnCl_2.2H_2O$	Tin chloride	[Sn] = 5.10^{-3}
CH_4N_5S	Thiourea	[S] = 40.10^{-3}

Table 1.
Chemical precursor concentrations used for the synthesis of CMTS thin film.

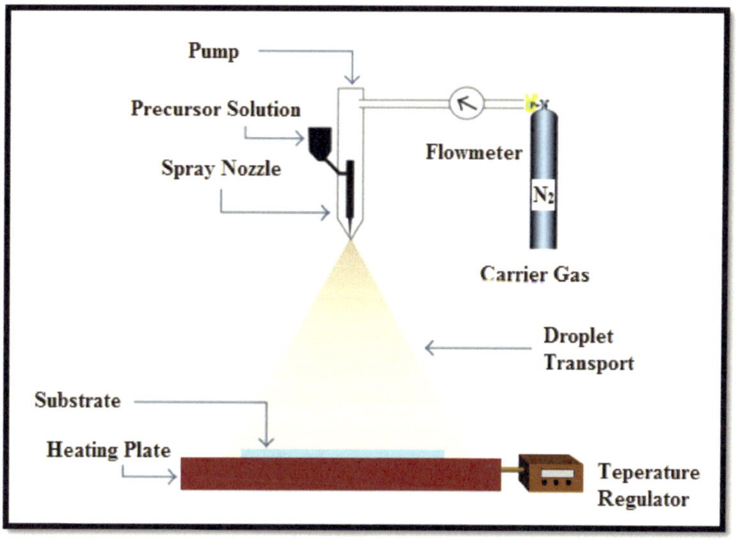

Figure 3.
Setup of the spray pyrolysis deposition of thin films.

compact, and porous film deposition at different temperatures. For example, in our recent work [7], a rice-like structure was obtained by preparing a CMTS thin film at Ts = 180°C, and uniformly covered films with a smooth, continuous, and dense surface were found at Ts = 210°C.

- It can use various substrates, like glass and metal oxides [36].
- It is easy to dope with various types of dopants in prepared solutions. Hammoud et al. [8] have grown a new Ag-doped CMTS thin film by SP.
- Spray Pyrolysis allows for the production of small and large surfaces; also, it is easy to obtain multilayers by SP. Nefzi et al. [36] synthesized an efficient p-type Cu_2FeSnS_4/n-type F-doped SnO_2.

4.5.2 Disadvantages

However, SP has several disadvantages:

- The formation of toxic gases.
- It is difficult to obtain a pure material without undesirable phases and contamination.
- Difficult to attend to exact substrate temperature because substrate temperature can be reduced by spraying precursors.
- Non-oxide films are complex to synthesize.

5. Case study: Substrate temperature effect

5.1 Structural properties

XRD patterns of the CMTS thin deposited at different substrate temperatures (Ts) are shown in **Figure 4**. The XRD patterns exhibit diffraction peaks located at 2θ = 28.7°, 32.5°, 47.4°, and 56.2° which can be assigned to the reticular CMTS planes (112), (200), (220), and (312), respectively, which are related to Cu_2MgSnS_4 Kesterite phase taking (112) reticular plane as preferred orientation (PDF 98–017–1983, ICDD) [11].

As shown in **Figure 4**, the peak intensities increase with the increase of substrate temperature. A significant increase in the (112) peak intensity is observed, followed by the appearance of a secondary phase, specifically Cu_2S, which decreases as the crystallinity of film improves, particularly for film deposited at Ts = 300°C. A similar trend has been reported in other works [7, 31, 37]. In our previous work, we investigated the impact of substrate temperature variations from 190 to 300°C on the physical properties of CMTS thin films prepared with an ethanolic solvent and found that Ts = 210°C was the optimal substrate temperature [7]. However, in this work, we use bi-distilled water as a solvent, which behaves differently. Water evaporates more slowly than alcohol, leaving the salts deposited on the substrate. Consequently, the resulting properties of the thin films are influenced by both the solvent type and its evaporation temperature. Since alcohol has a lower evaporation temperature than water, this can lead to a lower optimized substrate temperature for alcohol-based solutions compared to water-based ones. These XRD results indicate that the CMTS thin film grown at Ts = 300°C exhibits the best structural properties. Other studies on

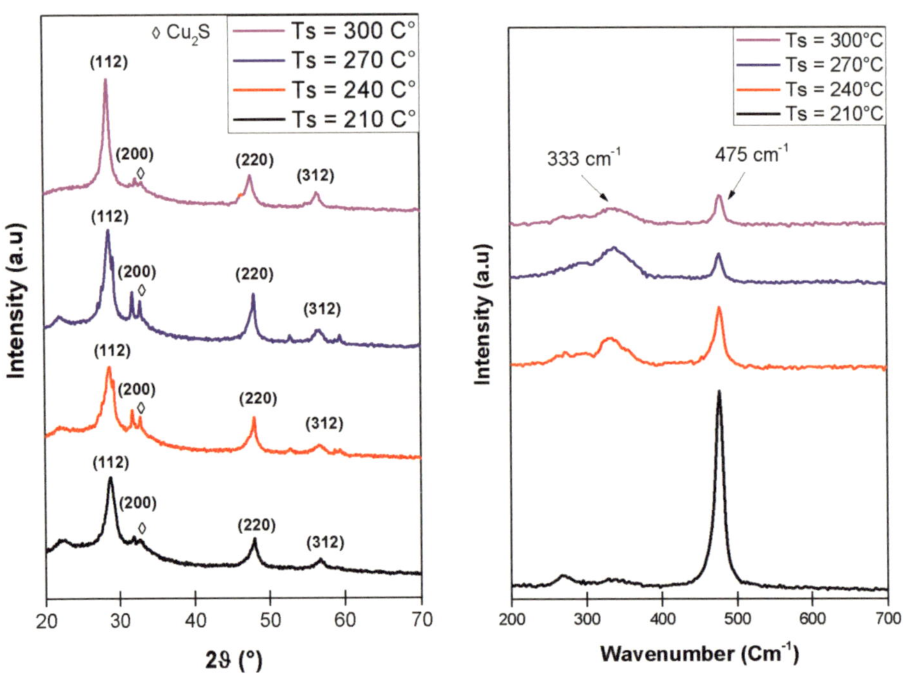

Figure 4.
XRD patterns and Raman spectra of CMTS films deposited at different substrate temperatures.

Quaternary compound	Optimal substrate temperatures (°C)
Cu_2ZnSnS_4 (CZTS)	280°C [39], 350°C [41]
Cu_2MgSnS_4 (CMTS)	210°C [7]
Cu_2FeSnS_4 (CFTS)	240°C [31]
Cu_2MnSnS_4 (CNTS)	360°C [40]

Table 2.
Summarizes the optimal substrate temperature for various quaternary compounds synthesized via spray pyrolysis [31, 39–41].

quaternary compounds prepared by spay pyrolysis have also shown that changes in Ts affect crystallinity of the films [31, 38].

Table 2 below summarizes the optimal substrate temperature achieved for quaternary compounds synthesized via spray pyrolysis.

To study the effect of the substrate temperature on the crystal structure, certain parameters such as crystallite size (D), dislocation density (δ), and strain (ε) were determined. The average crystallite size (D) was calculated from the Debye-Scherrer formula (Eq. (12)).

$$D = \frac{K\lambda}{\beta cos\theta} \tag{12}$$

where λ is the X-ray wavelength (1.5418 Å), K is the shape factor (K = 0.94), β is full width at half maximum (FWHM), and θ is the Bragg's diffraction angle.

The strain (ε) in the films was estimated from the following equation (Eq. (13)):

$$\varepsilon = \frac{\beta cos\theta}{4} \ (13) \tag{13}$$

The dislocation density (δ), which is defined as the length of dislocation lines per unit volume, is calculated using (Eq. (14)):

$$\delta = \frac{1}{D^2} \tag{14}$$

Table 3 presents the calculated crystallite size (D), the dislocation (δ), and the strain (ε) of CMTS thin films grown at different substrate temperatures. As shown,

Ts (°C)	2θ (°)	D (nm)	δ (10^{11} lines/m^2)	ε (10^{-3})
210	28.7	11.6	7.4	2.9
240	28.7	13.5	5.5	2.5
270	28.7	16.2	3.8	2.1
300	28.7	27.1	1.4	1.2

Table 3.
Crystallite size (D), dislocation density (δ), and strain (ε) calculated of the prepared CMTS thin film.

the crystallite size increases with the increase of substrate temperature, with the highest D value of approximately 27.1 nm obtained at Ts = 300°C. The dislocation density of CMTS films decreases as the substrate temperature increases, from 7.4×10^{11} lines/m^2 at Ts = 210°C to 2.5×10^{11} lines/m^2 at Ts = 300°C. These results reflect the improvement of crystallinity of the films with increasing substrate temperature. Additionally, the strain caused by stress in the films is at its lowest value of 1.2×10^{-3} for the sample deposited at Ts = 300°C. These results are consistent with the structural analysis, which indicates that Ts = 300°C yields the best crystalline quality for CMTS thin films.

Raman exhibits a total of 24 main vibrational modes in the Brillouin zone (Γ). These modes correspond to specific vibrational patterns within the crystal structure of CMTS, providing key insights into its phonon characteristics, which can be described by the following equation (Eq. (15)):

$$\Gamma = 3A + 7B + 7E \tag{15}$$

where A represents the non-degenerate modes, B refers to the singly degenerate modes, and E denotes the doubly degenerate modes.

Figure 4 shows the Raman spectra of CMTS thin films grown at various substrate temperatures, i.e. Ts = 210, 240, 270, and 300°C. It is evident that for all films, two main peaks appear at wave numbers 331 cm^{-1} and 474 cm^{-1}. Notably, the intensity of the peak at 331 cm^{-1} increases significantly as Ts increases from 210 to 300°C, while the intensity of the peak at 474 cm^{-1} decreases over the same temperature range. The peak at 331 cm^{-1}, which corresponds to the vibration of the sulfur atom in the CMTS phase, is associated with good crystallization, indicating that the CMTS thin film grown at Ts = 300°C has the best crystalline quality. These results align with previous works [2, 4, 5, 8, 10].

Additionally, we observe the presence of the Cu_2S vibrational mode at 474 cm^{-1} for all CMTS thin films grown at different Ts, confirming the coexistence of Cu_2S along with CMTS in the films [38, 42, 43]. These results are consistent with the XRD analysis.

5.2 Morphological studies

The substrate temperature plays a crucial role in determining the surface morphology of films synthesized by spray pyrolysis. **Figure 4** illustrates the effect of substrate temperature variation on the surface morphology of CMTS thin films. The SEM images in **Figure 5** clearly show that the surface of all films is uniformly covered with agglomerated CMTS particles, which is a characteristic commonly associated with the spray pyrolysis technique. As reported by Sekkati et al. [44] and Chen et al. [45], the substrate temperature is the primary factor influencing the film morphology. Also, they identified four distinct film morphologies: dense, dense with embedded particles, porous top layer with a dense bottom layer, and fractal-like porous structure.

At Ts = 210°C, the film exhibited a smooth but inhomogeneous surface, which can be attributed to the low substrate temperature. As the substrate temperature increased to 240°C, the morphology evolved, with the grains becoming more agglomerated, leading to a rougher surface with grain clusters. This suggests that higher temperatures promote grain growth and packing. For the sample grown at Ts = 270°C,

DOI: http://dx.doi.org/10.5772/intechopen.1009567

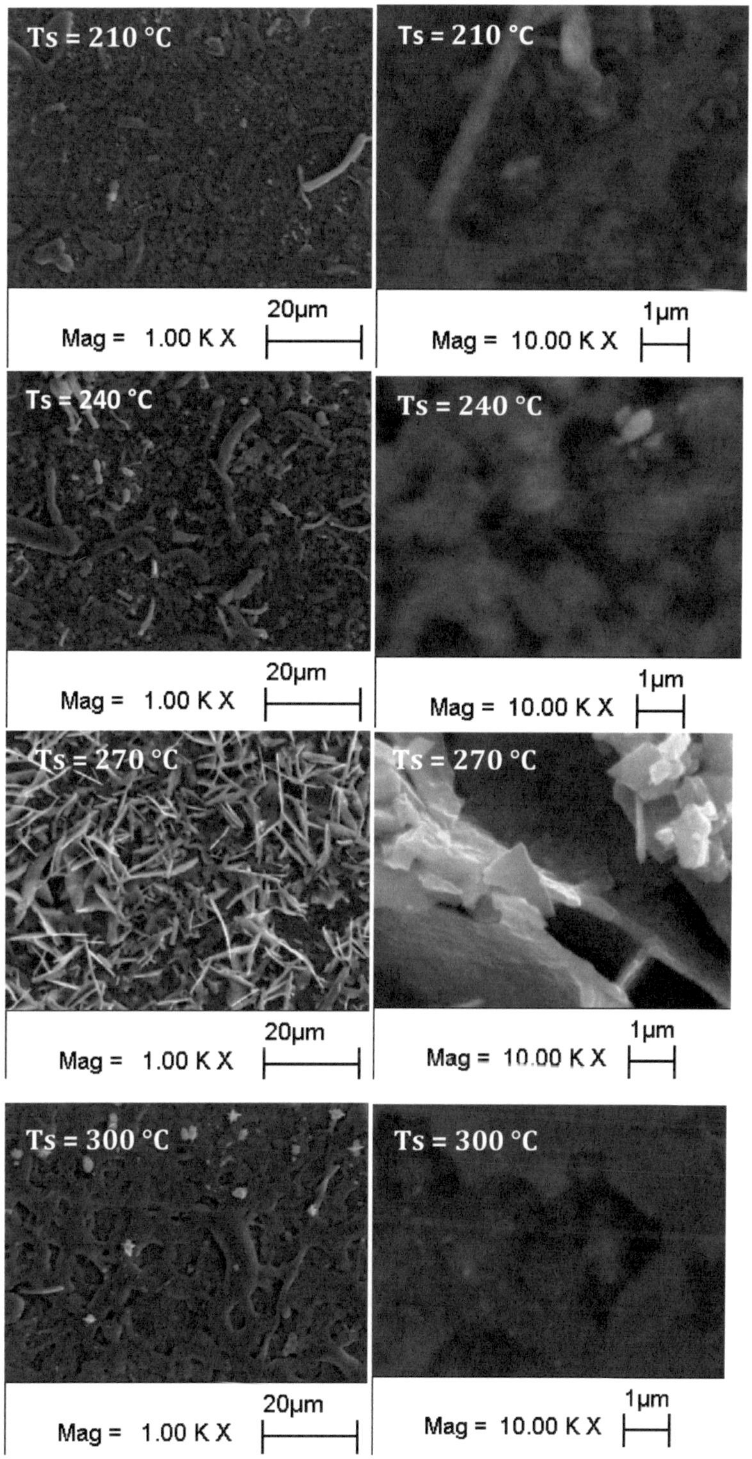

Figure 5.
SEM micrographs (two magnifications: 20 and 1 μm) of CMTS thin films synthesized at different substrate temperatures.

the surface took on a rice-like structure, with some areas showing nano-flakes or sheet-like structures, similar to the results observed by other researchers [8, 31]. This change may be related to a slightly Sn-rich composition, deviating from stoichiometry, likely due to the different reactivity of starting materials [46]. The same morphology has been reported in studies of Cu-rich compositions, such as those by K. G. et al. [43] and Tumbul et al. [47], when growing CZTS by ultrasonic spray pyrolysis. This could also explain the less crystalline nature observed, as mentioned in our previous work [7], where CMTS thin films were prepared by spray pyrolysis using ethanolic solvent at different substrate temperatures.

The sample grown at Ts = 300°C exhibited a clear interface between the substrate and the film, with no voids or cavities. The surface was uniformly covered with two distinct types of grains, forming a stacked-like structure. This indicates a more organized and compact grain arrangement compared to films deposited at lower substrate temperatures (210°C). The higher substrate temperature tended to promote greater grain growth, which was reflected in the larger grain size observed in the XRD analysis.

The development of different morphologies can be explained by considering the effect of the electric field and the droplet size, both of which are crucial for a range of applications. For example, highly porous materials are ideal for energy storage applications, while morphologies such as nanorods and nanotubes are better suited for solar cell applications [48].

5.3 Optical properties

The results clearly indicate that all the films exhibit low transmittance values, which can be divided into two regions: from 400 to 800 nm, where transmission remains below 1%, and from 850 to 1200 nm, where it increases to around 7% for all films. This suggests that the films have an absorption edge in the range of 800–850 nm. Furthermore, all CMTS samples showed low reflectance, with R remaining nearly constant as the wavelength increased, indicating the good quality of the CMTS thin film deposition. As observed in panel (d) of **Figure 5**, the CMTS films have high absorbance in the range [400–1200 nm], reaching up to 90%. These results are consistent with previous reports [10, 11], confirming the suitability of CMTS as an absorber material for photovoltaic applications [6].

The absorption coefficient (α) of CMTS thin films grown at various substrate temperatures is shown in **Figure 6**, calculated using the following equation (Eq. (16)):

$$\alpha = \frac{-1}{\mathrm{e}} \ln\left[\frac{T(\lambda)}{\left(1-R(\lambda)\right)^2}\right] \tag{16}$$

As seen, the absorption coefficient values for all CMTS films in the visible region range from 1.5×10^{4} to 4×10^{4} $(cm-1)$

The optical bandgap of the samples increases from 1.10 to 1.47 eV with increasing substrate temperature from 210 to 300°C (**Table 4**). More precisely, CMTS film grown at Ts = 300°C has a bandgap of 1.47 eV, making it a promising candidate for use as an absorbent layer in solar cells [6]. This increase in bandgap may be attributed to structural deformation, as shown in **Table 4**, where a decrease in dislocation density

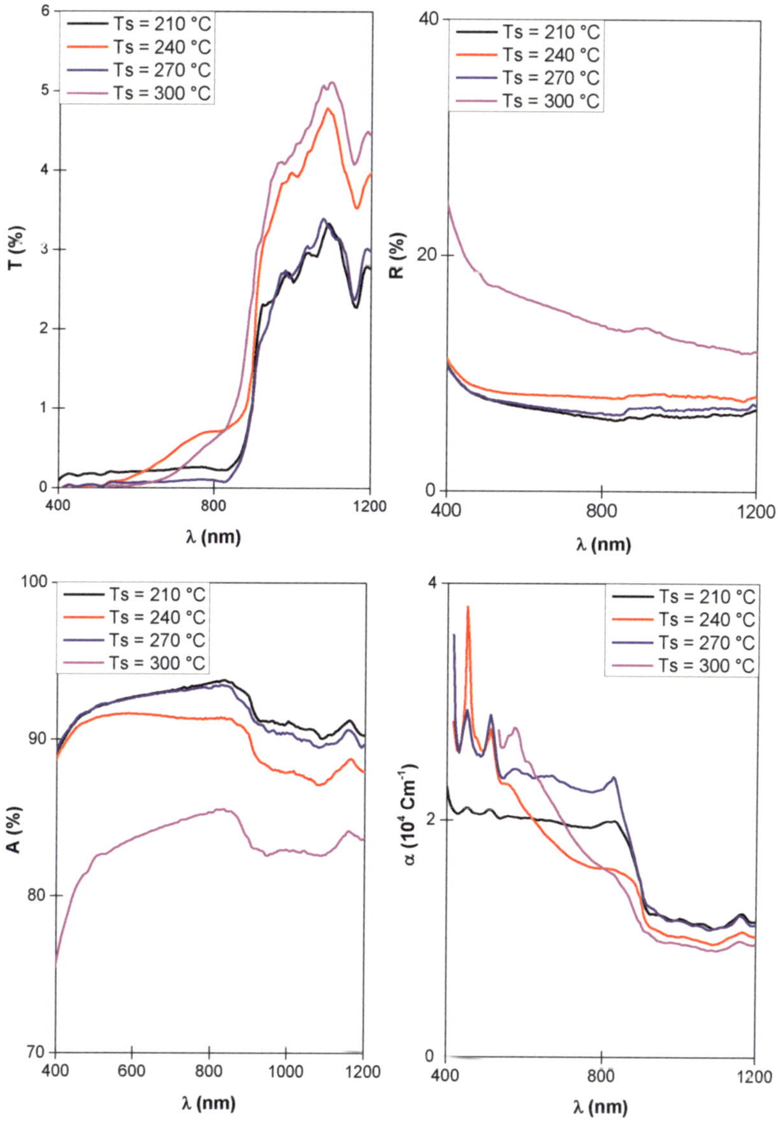

Figure 6.
Optical transmission (T), reflectance (R), absorbance (A), and absorption coefficient (α) spectra of CMTS obtained at various substrate temperatures.

Substrate temperature (Ts) (°C)	Gap energy (eV)
210	1.10
240	1.57
270	1.22
300	1.47

Table 4.
Optical bandgap energies of the CMTS thin films prepared at different substrate temperatures obtained from the Tauc relation.

is observed with higher substrate temperature. The optimal bandgap within the visible energy range, coupled with a high absorption coefficient, indicates that CMTS not only has potential for solar cell applications, but also for a variety of other opto-electronic uses [49].

6. Effect of doping: Ag doping

Doping is a common practice in spray pyrolysis to modify the properties of thin films [8]. Doping can be easily achieved in the SP technique by just adding small quantities of dopant to the precursor solution. Nguyen et al. [49] explored the effect of silver (Ag) doping on the physical properties and solar cell performance of Cu_2ZnSnS_4 film deposited by spray pyrolysis. Similarly, Hammoud et al. [8] observed that incorporating Ag into CMTS thin films prepared by spray pyrolysis improved their physical properties, particularly by reducing electrical resistivity due to an increase in electric charge carrier concentration.

Figure 7(a) presents the XRD patterns of 5% Ag-doped CMTS thin film, which clearly exhibits a kesterite structure with the (112) plane as the preferred orientation. Energy Dispersive X-ray analysis (EDAX), High-Resolution Transmission Electron Microscopy (HRTEM), and Selected Area Electron Diffraction (SAED) images of 5% Ag-doped CMTS deposited at Ts =300°C were shown in **Figure 7(b-d)**. EDAX analysis confirms the presence of all the expected elements in the Ag-doped CMTS sample. The HR-TEM image of the CMTS particle in **Figure 7(c)** shows an interplanar

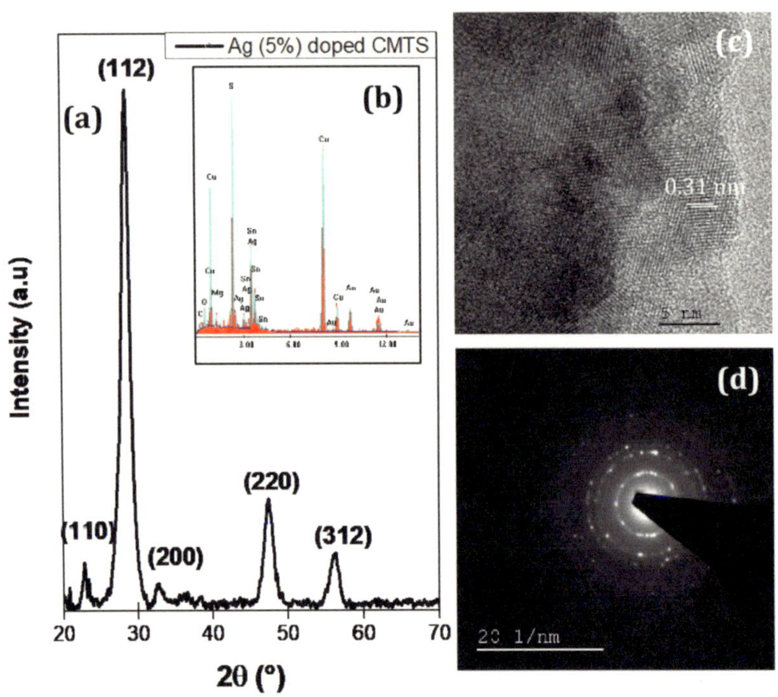

Figure 7.
XRD patterns, EDAX spectra, HR-TEM images, and SAED rings of the 5% Ag-doped CMTS nanoparticles (a–d).

spacing of approximately 3.6°A for the (112) plane. The SAED image in **Figure 7(b)** reveals the polycrystalline nature of CMTS particles, as indicated by the presence of diffraction spots corresponding to (112), (200), (220), and (312) reticular planes.

7. Application

The antibacterial activity of the 5% Ag-doped CMTS thin films was evaluated against *V. cholerae* and *S. Typhi* bacteria using zone of inhibition experiments, based on our previous work [8].

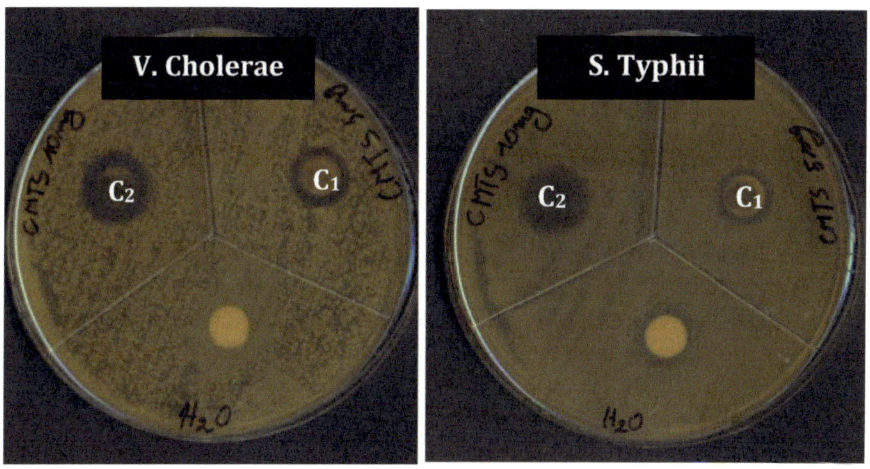

Figure 8.
Photographs of the antibacterial test against V. cholerae *and* S. Typhi *bacteria for 5% Ag-doped CMTS and negative control (bi-distilled).*

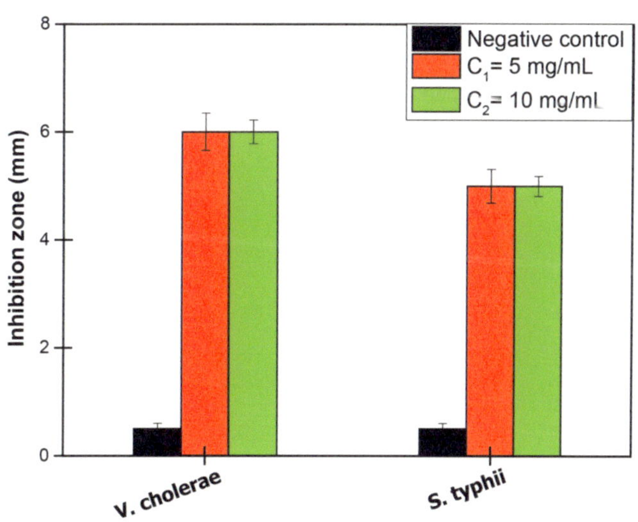

Figure 9.
Bar diagram corresponding to inhibition zone against V. cholerae *(red) and* S. Typhi *(green) bacteria for 5% Ag-doped CMTS and negative control (black).*

Figure 8 shows photographs of the antibacterial test against these bacteria using the agar-well diffusion method, for both the 5% Ag-doped CMTS and a negative control (bi-distilled). After the incubation period, larger zones of inhibition were observed for both bacteria, as the concentration increased from 5 to 10 mg/mL for 5% Ag-doped CMTS, demonstrating a rapid antibacterial effect. In contrast, the negative control, using bi-distilled water, showed no inhibition zone on the disk, indicating no antibacterial activity.

As shown in **Figure 9**, the inhibition zone diameter was approximately 6 mm for the 5% Ag-doped CMTS sample against *V. Cholerae* and 5 mm for the *S. Typhi* bacteria. These findings indicate that the highest antibacterial activity was observed for 5% Ag-doped CMTS against *V. Cholerae* at C_2 = 10 mg/mL, which is the highest concentration used in these tests. These results suggest that Ag-doped CMTS films hold great potential for antibacterial applications [8, 27].

8. Conclusion

To summarize, the description of an efficient elaboration technique "Spray Pyrolysis" of CMTS thin films and its deposition process were investigated. Moreover, the different characterization techniques used to study the structural, morphological, and optical properties of CMTS thin films were illustrated. In addition, an overview of the preparation conditions and doping impact on CMTS films was described. In the case of this study, the substrate temperature was varied from 210 to 300°C to explore its impact on the structural, morphological, and optical properties of CMTS thin films. The films deposited at 300°C exhibited excellent crystallinity, with a crystallite size of up to 27 nm, and favorable optical properties, such as a high absorption coefficient (3×10^4 cm^{-1}) and a band gap of 1.47 eV. These properties establish CMTS thin films as promising candidates for absorber layers in solar cells. Furthermore, the incorporation of 5% Ag as a dopant further improved both the structural and optical properties of CMTS thin films while introducing remarkable antibacterial functionality.

Conflict of interest

The authors declare no conflict of interest.

Author details

Amal Hammoud[1,2]*, Badriyah Alhalaili[3], Ruxandra Vidu[4] and Najoua Kamoun-Turki[2]

1 Department of Physics, Faculty of Sciences, Mentouri University of Constantine, Constantine, Algeria

2 LR99ES13 Physic of Condensed Matter Laboratory (PCML), Department of Physics, Faculty of Sciences of Tunis El Manar, Tunis, Tunisia

3 Nanotechnology and Advanced Materials Program, Kuwait Institute for Scientific Research, State of Kuwait

4 Faculty of Materials Science and Engineer, National University of Science and Technology, POLITEHNICA, Bucharest, Romania

*Address all correspondence to: amal.hammoud@umc.edu.dz

References

[1] Bekki B, Amara K, El Keurti M. First-principles study of the new potential photovoltaic absorber: Cu_2MgSnS_4 compound. Chinese Physics B. 2017;**26**(7):076201. DOI: 10.1088/1674-1056/26/7/076201

[2] Wei M, Du Q, Wang R, Jiang G, Liu W, Zhu C. Synthesis of new earth-abundant Kesterite Cu_2MgSnS_4 nanoparticles by hot-injection method. Chemistry Letters. 2014;**43**(7):1149-1151. DOI: 10.1246/cl.140208

[3] Guo Y. The structural, morphological, and optical–electrical characteristics of Cu_2XSnS_4 (X = Cu, Mg) thin films fabricated by novel ultrasonic co-spray pyrolysis. Materials Letters. 2016;**172**:4

[4] Sharma A, Sahoo P, Singha A, Padhan S, Udayabhanu G, Thangavel R. Efficient visible-light-driven water splitting performance of sulfidation-free, solution processed Cu_2MgSnS_4 thin films: Role of post-drying temperature. Solar Energy. 2020;**203**:284-295. DOI: 10.1016/j.solener.2020.04.027

[5] Ali A, Syed WA, Arif W, Ahmed S, Ghufran MA. Environment friendly nanoparticles of quaternary compound $Cu_{0.5}Mg_xSnS_4$ for possible photovoltaic and photo-catalytic applications. Materials Research Express. 2019;**6**(9):0950b5. DOI: 10.1088/2053-1591/ab0f7e

[6] Yang G et al. Synthesis and characterizations of Cu_2MgSnS_4 thin films with different sulfuration temperatures. Materials Letters. 2019;**242**:58-61. DOI: 10.1016/j.matlet.2019.01.102

[7] Hammoud A, Souli M, Kouki F, Ajili L, Bouricha B, Kamoun N. Investigation of substrate temperature effect on physical properties of sprayed Cu_2MgSnS_4 thin films for solar cells and humidity sensing applications. Journal of Materials Science: Materials in Electronics. 2022;**127**:112296. DOI: 10.1007/s10854-022-07872-z

[8] Hammoud A, Souli M, Diouani MF, Alhalaili B, Vidu R, Kamoun-Turki N. Synthesis and characterization of novel sprayed Ag-doped quaternary Cu_2MgSnS_4 thin film for antibacterial application. Nanomaterials. 2022;**12**(19):3459. DOI: 10.3390/nano12193459

[9] Ali A, Liang Y, Ahmed S, Yang B, Guo B, Yang Y. Mutual contaminants relational realization and photocatalytic treatment using Cu_2MgSnS_4 decorated BaTiO3. Applied Materials Today. 2020;**18**:100534. DOI: 10.1016/j.apmt.2019.100534

[10] Hammoud A et al. Investigation on Cu_2MgSnS_4 thin film prepared by spray pyrolysis for photovoltaic and humidity sensor applications. Optical Materials. 2022;**127**:112296. DOI: 10.1016/j.optmat.2022.112296

[11] Hammoud A, Yahmadi B, Souli M, Ahmed SA, Ajili L, Kamoun-Turki N. Effect of sulfur content on improving physical properties of new sprayed Cu_2MgSnS_4 thin films compound for optoelectronic applications. European Physical Journal Plus. 2022;**137**(2):232. DOI: 10.1140/epjp/s13360-022-02417-z

[12] Imtiaz F, Rashid J, Xu M. Semiconductor nanocomposites for visible light photocatalysis of water pollutants. In: Rahman M, Khan A, Asiri A, Inamuddin I, editors. Concepts of Semiconductor Photocatalysis.

London, UK: IntechOpen; 2019. DOI: 10.5772/intechopen.86542

[13] Shwetharani R, Chandan HR, Sakar M, Balakrishna GR, Reddy KR, Raghu AV. Photocatalytic semiconductor thin films for hydrogen production and environmental applications. International Journal of Hydrogen Energy. 2020;**45**(36):18289-18308. DOI: 10.1016/j.ijhydene.2019.03.149

[14] Liu Q et al. 18% efficiency organic solar cells. Science Bulletin. 2020;**65**(4):272-275. DOI: 10.1016/j.scib.2020.01.001

[15] Vozniuk O, Tanchoux N, Millet J-M, Albonetti S, Di Renzo F, Cavani F. Spinel mixed oxides for chemical-loop reforming: From solid state to potential application. In: Studies in Surface Science and Catalysis. Amsterdam, Netherlands: Elsevier; 2019;**178**:281-302. DOI: 10.1016/B978-0-444-64127-4.00014-8

[16] Patil MP, Kim G-D. Eco-friendly approach for nanoparticles synthesis and mechanism behind antibacterial activity of silver and anticancer activity of gold nanoparticles. Applied Microbiology and Biotechnology. 2017;**101**(1):79-92. DOI: 10.1007/s00253-016-8012-8

[17] Bijelic A, Rompel A. Polyoxometalates: More than a phasing tool in protein crystallography. ChemTexts. 2018;**4**(3):10. DOI: 10.1007/s40828-018-0064-1

[18] Ferrari AC. Raman spectroscopy of graphene and graphite: Disorder, electron–phonon coupling, doping and nonadiabatic effects. Solid State Communications. 2007;**143**(1-2):47-57. DOI: 10.1016/j.ssc.2007.03.052

[19] Sun D, Siddiqui MOR, Iqbal K. Specialty testing techniques for smart textiles. In: Smith WC, editor. Smart Textile Coatings and Laminates. 2nd ed. Cambridge, UK: Woodhead Publishing; 2019. pp. 99-116. DOI: 10.1016/B978-0-08-102428-7.00004-3

[20] Nasir S, Hussein MZ, Zainal Z, Yusof NA, Mohd Zobir SA, Alibe IM. Potential valorization of by-product materials from oil palm: A review of alternative and sustainable carbon sources for carbon-based nanomaterials synthesis. BioResources. 2019;**14**(1):2352-2388. DOI: 10.15376/biores.14.1.Nasir

[21] Ayodhya D, Venkatesham M, Kumari AS, Reddy GB, Ramakrishna D, Veerabhadram G. Photocatalytic degradation of dye pollutants under solar, visible and UV lights using green synthesised CuS nanoparticles. Journal of Experimental Nanoscience. 2016;**11**(6):418-432. DOI: 10.1080/17458080.2015.1070312

[22] Reddy VRM. Review on Cu_2SnS_3, Cu_3SnS_4, and Cu_4SnS_4 thin films and their photovoltaic performance. Journal of Industrial and Engineering Chemistry. 2019;**76**:36. DOI: 10.1016/j.jiec.2019.03.035

[23] Chavda A, Patel B, Marathey P, Mukhopadhyay I, Ray A. Articulating effect of low copper content on structure and optoelectronic properties of spray deposited Cu_2ZnSnS_4 thin films – From experiment and first-principles investigations. Materials Science and Engineering: B. 2021;**263**:114912. DOI: 10.1016/j.mseb.2020.114912

[24] Huang TJ, Zhang L, Shi D, Qi G, Gong H. Dual use of Cu_2ZnSnS_4 in solar cells and energy storage devices. Solar Energy Materials and Solar Cells. 2018;**180**:328-333. DOI: 10.1016/j.solmat.2017.07.003

[25] Sankapal BR, Ennaoui A, Gupta RB, Lokhande CD, . Simple

Chemical Methods for Thin Film Deposition: Synthesis and Applications. Singapore: Springer Nature Singapore; 2023. DOI: 10.1007/978-981-99-0961-2

[26] Velumani M, Meher SR, Alex ZC. Impedometric humidity sensing characteristics of SnO2 thin films and SnO_2–ZnO composite thin films grown by magnetron sputtering. Journal of Materials Science: Materials in Electronics. 2018;**29**(5):3999-4010. DOI: 10.1007/s10854-017-8342-z

[27] Agawane GL, Vanalakar SA, Kamble AS, Moholkar AV, Kim JH. Fabrication of $Cu_2(Zn_xMg_{1-x})SnS_4$ thin films by pulsed laser deposition technique for solar cell applications. Materials Science in Semiconductor Processing. 2018;**76**:50-54. DOI: 10.1016/j.mssp.2017.12.010

[28] Laghfour Z et al. Sodium doping mechanism on sol-gel processed kesterite Cu_2ZnSnS_4 thin films. Superlattices and Microstructures. 2018;**120**:747-752. DOI: 10.1016/j.spmi.2018.05.018

[29] Ziti A. Characteristics of kesterite CZTS thin films deposited by dip-coating technique for solar cell applications. Journal of Materials Science. 2019;**14**:10. DOI: 10.1007/s10854-019-01676-4

[30] Ziti A et al. Study of kesterite CZTS thin films deposited by spin coating technique for photovoltaic applications. Superlattices and Microstructures. 2019;**127**:191-200. DOI: 10.1016/j.spmi.2017.11.023

[31] Nefzi C, Souli M, Cuminal Y, Kamoun-Turki N. Effect of substrate temperature on physical properties of Cu_2FeSnS_4 thin films for photocatalysis applications. Materials Science and Engineering: B. 2020;**254**:114509. DOI: 10.1016/j.mseb.2020.114509

[32] Souli M, Engazou R, Ajili L, Kamoun-Turki N. Physical properties evolution of sprayed Cu_2MgSnS_4 thin films with growth parameters and vacuum annealing. Superlattices and Microstructures. 2020;**147**:106711. DOI: 10.1016/j.spmi.2020.106711

[33] Chen C, Kelder EM, Van Der Put PJJM, Schoonman J. Morphology control of thin $LiCoO_2$ films fabricated using the electrostatic spray deposition (ESD) technique. Journal of Materials Chemistry. 1996;**6**(5):765. DOI: 10.1039/jm9960600765

[34] Gunavathy KV. Effect of solvent on the characteristic properties of nebulizer spray pyrolyzed Cu_2ZnSnS_4 absorber thin films for photovoltaic application. Thin Solid Films. 2020;**697**:10. DOI: 10.1016/j.tsf.2020.137831

[35] Ajili M, Castagné M, Kamoun Turki N. Spray solution flow rate effect on growth, optoelectronic characteristics and photoluminescence of SnO_2:F thin films for photovoltaic application. Optik. 2015;**126**(7-8):708-714. DOI: 10.1016/j.ijleo.2015.02.039

[36] Nefzi C, Souli M, Jeyadevan B, Kamoun-Turki N. Effect of sprayed volume on physical properties of Cu_2FeSnS_4 thin films and an efficient p-type Cu_2FeSnS_4/n-type F-doped SnO_2 heterojunction. Journal of Physics and Chemistry of Solids. 2020;**144**:109497. DOI: 10.1016/j.jpcs.2020.109497

[37] Bakr NA, Khodair ZT, Hassan SSA. Effect of Substrate Temperature on Structural and Optical Properties of Cu_2ZnSnS_4 (CZTS) Films Prepared by Chemical Spray Pyrolysis Method, *Undefined*. 2015. Available from: https://www.semanticscholar.org/paper/Effect-of-Substrate-Temperature-on-Structural-and-2-Bakr-Khodair/ffc7fe4a6801129e35e0b30b55b6599c9c731389 [Accessed: July 17, 2021]

[38] Arba Y. Effect of Substrate Temperature on Physical Properties of CZTS Thin Films. 2015. doi: 10.13140/RG.2.1.1974.9845.

[39] Seboui Z, Gassoumi A, Cuminal Y, Kamoun-Turki N. Effect of annealing process on the properties of Cu_2ZnSnS_4 thin films. Superlattices and Microstructures. 2014;**75**:586-592. DOI: 10.1016/j.spmi.2014.07.025

[40] Nie L, Yang J, Yang D, Liu S. Effect of substrate temperature on growth and properties of Cu_2MnSnS_4 thin films prepared by chemical spray pyrolysis. Journal of Materials Science: Materials in Electronics. 2019;**30**(4):3760-3766. DOI: 10.1007/s10854-018-00658-2

[41] Khalate SA, Kate RS, Kim JH, Pawar SM, Deokate RJ. Effect of deposition temperature on the properties of Cu_2ZnSnS_4 (CZTS) thin films. Superlattices and Microstructures. 2017;**103**:335-342. DOI: 10.1016/j.spmi.2017.02.003

[42] Seboui Z, Cuminal Y, Kamoun-Turki N. Physical properties of Cu_2ZnSnS_4 thin films deposited by spray pyrolysis technique. Journal of Renewable and Sustainable Energy. 2013;**5**(2):023113. DOI: 10.1063/1.4795399

[43] Deepa KG, Chandrabose G, Jampana N. Tailoring the properties of Cu_2ZnSnS_4 thin films grown by ultrasonic spray pyrolysis. Journal of Analytical and Applied Pyrolysis. 2016;**120**:356-360. DOI: 10.1016/j.jaap.2016.06.003

[44] Sekkati M, Taibi M, Schmerber G, Benamar EB, Regragui M, Cherkaoui El Moursli F, et al. The impact of Na and K on properties of Cu_2ZnSnS_4 thin films prepared by ultrasonic spray technique. Physica Status Solidi C. 2017;**14**:1700170. DOI: 10.1002/pssc.201700170

[45] Chen L et al. Influence of annealing temperature on structural and optical properties of Cu2MnSnS4 thin films fabricated by sol–gel technique. Journal of Alloys and Compounds. 2015;**640**:23-28. DOI: 10.1016/j.jallcom.2015.03.225

[46] Long F, Mo S, Zeng Y, Chi S, Zou Z. Synthesis of flower-like Cu_2ZnSnS_4 Nanoflakes via a microwave-assisted Solvothermal route. International Journal of Photoenergy. 2014;**2014**:1-4. DOI: 10.1155/2014/618789

[47] Tumbul A, Göktaş A, Zarbali MZ, Aslan F. Structural, morphological and optical properties of the vacuum-free processed CZTS thin film absorbers. Materials Research Express. 2018;**5**(6):066408. DOI: 10.1088/2053-1591/aac80e

[48] Wan H, Peng G, Yao X, Yang J, Cui P, Xu X. Cu_2ZnSnS_4/graphene nanocomposites for ultrafast, long life all-solid-state lithium batteries using lithium metal anode. Energy Storage Materials. 2016;**4**:59-65. DOI: 10.1016/j.ensm.2016.02.004

[49] Nguyen TH et al. Structural and solar cell properties of a Ag-containing $Cu_2 ZnSnS_4$ thin film derived from spray pyrolysis. ACS Applied Materials and Interfaces. 2018;**10**(6):5455-5463. DOI: 10.1021/acsami.7b14929

Section 2

Relevant Technological Applications

Chapter 4

Application of Thin Films in High-Voltage and Power Electronics

Nafees Ahmad

Abstract

Thin films, ranging in thickness from a few nanometers to several micrometers, play a key role in the development of high-voltage and power electronics technologies. Using precise deposition techniques, these ultra-thin layers provide essential properties such as insulation, thermal management, and improved electrical performance. Using thin films, device miniaturization and performance have been significantly improved, allowing them to be integrated into advanced applications such as capacitors, resistors, and transistors. Their role is particularly evident in wide-bandgap semiconductor devices such as silicon carbide (SiC) and gallium nitride (GaN), which have excellent switching speed, breakdown voltage, and thermal performance. This chapter highlights the importance of advanced deposition methods, including atomic layer deposition (ALD) and chemical vapor deposition (CVD), which ensure precise control of film thickness and structure. These technologies produce application-specific, defect-free layers, which are critical for maintaining the reliability and performance of high-voltage systems. Additionally, a variety of materials used in the films, such as high-k dielectrics and nanocomposites, facilitate the development of high capacitance density capacitors and high-performance resistors. Their use in real-world scenarios such as electric vehicles and renewable energy systems has advanced power electronics technology, improving energy conversion, thermal stability, and overall efficiency. Going forward, the integration of new materials and emerging technologies is expected to further enhance the performance of the films. Research into two-dimensional materials such as graphene and transition metal dichalcogenides, as well as nanostructured films embedded with quantum dots and nanowires, offers the potential for next-generation energy storage and conversion devices. Furthermore, flexible and stretchable films open the way to wearable electronics and medical devices. By solving challenges such as thermal management and scalability, thin films have the potential to revolutionize power electronics, driving innovation in applications ranging from smart grids to advanced aerospace systems.

Keywords: thin films, high-voltage electronics, power electronics, miniaturization, atomic layer deposition (ALD), chemical vapor deposition (CVD)

1. Introduction

1.1 Overview of thin films in high-voltage and power electronics

Thin films, defined by their nanoscale to microscale thicknesses, have become indispensable in numerous advanced technologies, including high-voltage and power electronics. Due to their nanometer-scale thickness and highly tunable properties, thin films have revolutionized high-voltage and power electronics technology, becoming an indispensable class of materials in modern engineering. These films, which typically range in thickness from a few nanometers to several micrometers, are deposited on substrates to improve specific properties such as electrical insulation, thermal management, and mechanical stability. The emergence of thin film technology has not only made significant progress in the miniaturization of electronic devices but also improved their performance and efficiency under extreme operating conditions [1]. Unlike traditional bulk materials, thin films allow precise control of the composition and structure of the material, which is important for meeting the stringent requirements of high-voltage and power systems [2].

High-voltage electronics typically operate at voltages above 1 kV, while power electronics are designed for efficient power conversion and control and are the foundation of modern energy systems [3]. These systems require materials that can withstand high electric fields, exhibit excellent thermal conductivity, and maintain reliability over long periods of stress. Thin films have emerged as a solution to these challenges because of their ability to act as dielectric layers, conductive paths, or protective barriers, depending on the application. For example, wide bandgap semiconductors, such as silicon carbide (SiC) and gallium nitride (GaN), have attracted considerable attention as thin film materials due to their excellent electrical and thermal properties. These materials enable advances in power density and switching speed, making them indispensable for next-generation electronic devices [4].

The incorporation of nanofillers into thin film materials has further augmented their performance, particularly in mitigating critical challenges such as partial discharge, dv/dt stresses, and space charge accumulation [5]. Nanofillers enhance the dielectric properties of thin films by reducing electrical conductivity, suppressing charge injection, and increasing breakdown strength, making them highly effective for high-voltage insulation applications [6]. Furthermore, advancements in semiconductor packaging have leveraged thin film technologies to address issues like thermal management, mechanical stress, and the prevention of triple point phenomena, which occur at material interfaces and are a major cause of electrical failure [7]. These improvements have been instrumental in combating the aging and degradation mechanisms commonly observed in power electronic devices operating under extreme conditions. By enabling more robust and reliable device performance, thin films with nanofillers play a pivotal role in extending the operational life of high-voltage and power electronic systems.

The production of thin films involves a series of advanced deposition techniques, each developed to achieve specific material properties. Physical vapor deposition (PVD), chemical vapor deposition (CVD), and atomic layer deposition (ALD) are among the most prominent methods that allow precise control of film thickness, structure, and uniformity [8]. Techniques such as ALD, with its layer-by-layer growth mechanism, have proven particularly beneficial for applications that require precision on the nanometer scale. Combining these advanced deposition technologies with new materials not only improves device performance but also expands the scope of thin film applications. From high-voltage capacitors and resistors to thermal management solutions and optoelectronic components, the strategic use of thin films plays a significant role in overcoming the limitations of traditional materials, paving the way for innovation in both established and emerging technologies.

This chapter provides a comprehensive overview of the role of thin films in high-voltage and power electronics, focusing on their material properties, deposition techniques, and practical applications. Recent advances in this field are highlighted, including the use of wide bandgap materials and precision deposition techniques that have played a significant role in the development of high-performance electronic devices. Through this detailed exploration, this chapter aims to bridge the gap between basic research and industrial applications, highlighting the transformative potential of thin films in addressing contemporary engineering challenges.

1.2 Importance and relevance of thin films

Thin film is a very crucial technology for high-voltage and power electronics. With electronic device footprints shrinking and functionality increasing, technology has exploded in demand for materials able to withstand the higher electrical stresses along with the thermal load. These needs were met by thin films, which provided better electrical isolation, heightened thermal conductivity, and supported high-power densities.

Thin films also find application in the fabrication of wide bandgap semiconductor devices. These devices, often manufactured from SiC and GaN material, have advantages and early-on disadvantages in the form of thin film technologies that can make them more efficient while improving thermal management capabilities. Given its superior properties, encompassing higher breakdown voltages and faster switching speeds compared to established semiconductor materials, wide bandgap material is perfect for high-voltage and power applications. The utility of these materials, however, is unlocked by using thin films and providing electrical/thermal interfaces.

Thin films are also important for high-voltage insulation purposes. Advanced thin film insulators are more frequently being used instead of, or in conjunction with, traditional materials such as porcelain and glass. Thin films have high dielectric strength and show superior performance in extreme conditions required for the operation of explosion-proof systems with a power supply applying approximately 1 kV, safe enough under all circumstances [9].

1.3 Objectives and scope of the chapter

This chapter aims to give an exhaustive description of various high-voltage and power electronic applications using thin films. It covers a comprehensive review of the wide variety of materials and deposition methods that are used in thin film processing, along with specific applications to capacitors, variable frequency resistors, high-leverage semiconductor devices, and insulation under steep voltage. Recent advances in thin film technology on high-voltage and power electronic devices, as well as expert opinions, are introduced in this chapter.

2. Fundamentals of thin films

2.1 Basic properties of thin films

Thin films are layers of material ranging in thickness from a few nanometers to several micrometers. Their basic properties, which include mechanical, electrical, optical, and thermal characteristics, are highly dependent on their composition, structure, and the deposition process used. One of the critical properties of thin films is their surface-to-volume ratio, which significantly impacts their behavior compared to bulk materials. This high surface area can enhance reactivity and influence the electrical and thermal conductivities of the films.

Thermal properties, such as thermal conductivity and expansion, play a vital role in managing heat dissipation in high-power electronic devices. Advanced thin film materials like diamond-like carbon (DLC) and graphene have been developed to provide superior thermal management due to their high thermal conductivity [9]. The ongoing research and development in nanotechnology and materials science continue to enhance these properties, leading to more efficient and reliable thin film applications in high-voltage and power electronics.

2.2 Classification and types of thin films

Thin films can be classified based on their materials, structure, and application methods. Material-wise, thin films are typically categorized into metallic, dielectric, and semiconductor films. Metallic thin films, such as those made from aluminum, gold, and copper, are widely used for their excellent electrical conductivity and are crucial in interconnects and electrodes. Dielectric thin films, including silicon dioxide (SiO_2) and titanium dioxide (TiO_2), are essential for their insulating properties in capacitors and gate dielectrics. Semiconductor thin films, such as silicon (Si) and gallium nitride (GaN), are the backbone of modern electronic devices, including transistors and diodes [10].

Based on the application methods, thin films are classified into physical vapor deposition (PVD), chemical vapor deposition (CVD), and atomic layer deposition (ALD), among others. PVD methods, such as sputtering and evaporation, involve the physical transfer of material from a source to the substrate [11]. CVD techniques use chemical reactions to deposit materials from gaseous precursors, offering high conformity and uniformity. ALD, a subclass of CVD, allows for atomic-scale precision in film thickness, making it ideal for applications requiring ultra-thin and highly uniform films [12].

2.3 Material properties relevant to high-voltage and power applications

The material properties of thin films that are particularly relevant to high-voltage and power applications include dielectric strength, thermal conductivity, and electrical resistivity. Dielectric strength is a measure of a material's ability to withstand electric fields without breaking down, which is critical for insulators used in high-voltage devices. Materials such as silicon nitride (Si_3N_4) and aluminum oxide (Al_2O_3) are commonly used for their high dielectric strengths, enabling reliable operation in high-voltage environments [13].

Recent advancements have led to the development of novel materials with enhanced properties. For instance, graphene and transition metal dichalcogenides

(TMDs) have shown promise in achieving superior electrical and thermal performance due to their unique two-dimensional structures and high carrier mobilities. These advancements are paving the way for more efficient and compact high-voltage and power electronic devices.

3. Deposition methods for thin films

3.1 Overview of deposition methods

The deposition methods for thin films are critical in determining their structural, mechanical, electrical, and optical properties. The choice of deposition technique depends on the desired film characteristics, substrate material, and application. The primary methods of thin film deposition include Physical Vapor Deposition (PVD), Chemical Vapor Deposition (CVD), and Atomic Layer Deposition (ALD), each offering unique advantages and specific application areas. Additionally, other methods like sol-gel processing, sputtering, and electrochemical deposition are employed for specific requirements.

3.2 Physical vapor deposition (PVD)

Physical Vapor Deposition (PVD) involves the physical transfer of material from a source to a substrate in a vacuum environment. The PVD process includes several techniques, such as evaporation, sputtering, and pulsed laser deposition. During evaporation, the material is heated until it vaporizes and then condenses onto the substrate. Sputtering involves the ejection of material from a target by bombarding it with high-energy ions, which then deposit onto the substrate. The schematic illustration of the physical vapor deposition process is given in **Figure 1**.

PVD techniques offer high purity and excellent control over film thickness and composition. They are widely used for depositing metals, dielectrics, and

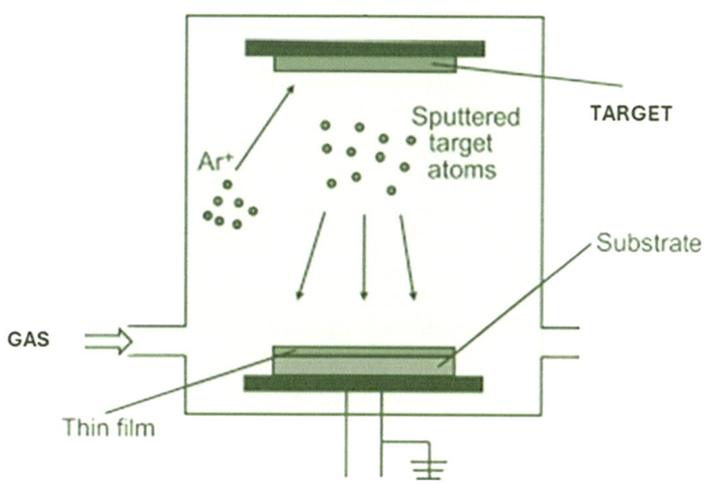

Figure 1.
Schematic illustration of the physical vapor deposition process.

semiconductors. However, PVD requires a high vacuum environment and can be limited by the line-of-sight deposition, which may not be suitable for complex geometries.

3.3 Chemical vapor deposition (CVD)

Chemical Vapor Deposition (CVD) involves the chemical reaction of gaseous precursors to form a solid material on a substrate. The process can be conducted at atmospheric or reduced pressures. Variations of CVD include plasma-enhanced CVD (PECVD), which uses plasma to enhance chemical reactions, and low-pressure CVD (LPCVD), which operates at reduced pressures to improve film uniformity and conformality. The schematic illustration of the chemical vapor deposition process is given in **Figure 2**.

CVD provides excellent conformity and uniformity over large areas and complex shapes, making it suitable for high-aspect-ratio structures. It allows for the deposition of a wide range of materials, including metals, insulators, and semiconductors. However, CVD processes can be complex and require precise control of reaction parameters. They also involve elevated temperatures, which may not be compatible with all substrates.

3.4 Atomic layer deposition (ALD)

Atomic Layer Deposition (ALD) is a highly controlled technique that deposits thin films one atomic layer at a time. ALD involves alternating exposure of the substrate to different precursors, with each exposure resulting in a self-limiting reaction that deposits a single atomic layer. This process allows for atomic-scale precision in film thickness and composition. The schematic illustration of the atomic layer deposition process is given in **Figure 3**.

ALD offers unparalleled control over film thickness and composition, making it ideal for applications requiring ultra-thin and highly uniform films. It is particularly useful for depositing high-k dielectrics, metal oxides, and nitrides. However, ALD processes are typically slower than other deposition methods and can be more complex and expensive.

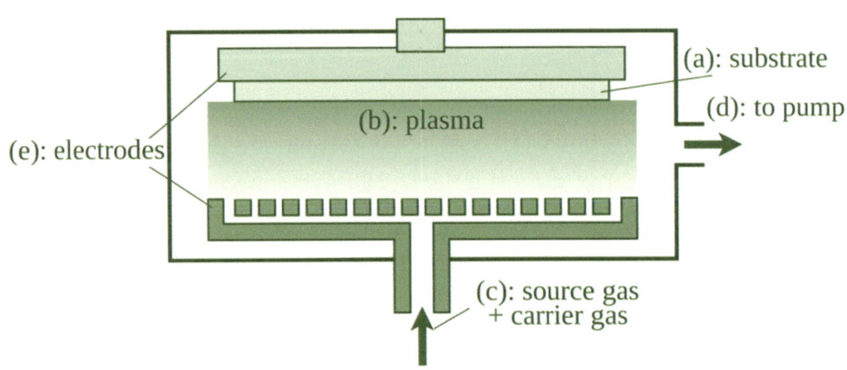

Figure 2.
Schematic illustration of the chemical vapor deposition process.

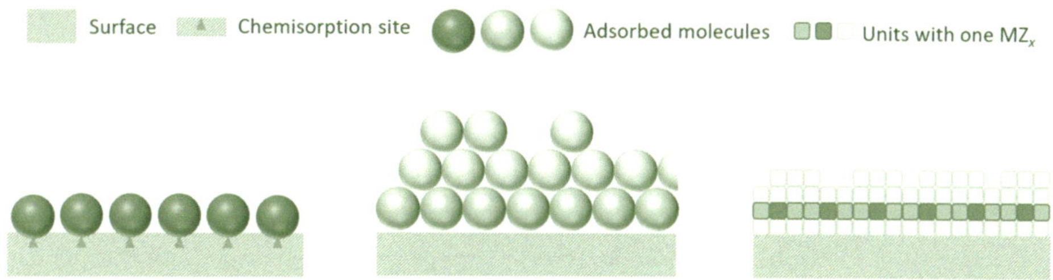

Figure 3.
Schematic illustration of the atomic layer deposition process.

3.5 Other deposition methods

i. The sol-gel process involves the transition of a solution (sol) into a solid (gel) to form thin films. This method is widely used for depositing ceramic and glass materials. The sol-gel process is cost-effective and can produce films with excellent optical and electrical properties. However, it may require post-deposition annealing to improve film quality.

ii. Sputtering, a PVD technique, involves the ejection of material from a target by bombarding it with high-energy ions. The ejected material then deposits onto the substrate. Sputtering is versatile and can deposit a wide range of materials, including metals, alloys, and dielectrics. It offers good adhesion and uniformity but may have lower deposition rates compared to other methods.

iii. Electrochemical deposition involves the reduction of metal ions in an electrolyte solution onto a conductive substrate. This technique is used for depositing metals and alloys with precise control over thickness and composition. Electrochemical deposition is cost-effective and scalable, but it is limited to conductive substrates and specific materials.

4. Different types of WBG materials

Wide Band Gap (WBG) materials are critical for advancing high-voltage and power electronic applications due to their superior electrical, thermal, and mechanical properties. These materials are categorized based on their functionality into conductive, insulating, and semiconducting types. Each category plays a vital role in enhancing device performance, efficiency, and reliability under demanding conditions.

4.1 Conductive materials

Conductive materials form the backbone of electrical interconnects and electrodes in high-voltage devices. Among these, aluminum is widely used due to its lightweight nature, good electrical conductivity, and cost-effectiveness. It is a preferred material in capacitors and interconnects where economic and practical factors are important. Copper, known for its excellent electrical and thermal conductivity, is often employed in high-power and high-frequency applications, providing superior heat dissipation for high-performance devices. Precious metals like gold and silver are also utilized in critical applications due to their excellent electrical properties. Gold's resistance to oxidation and stable performance under extreme conditions make it suitable for high-reliability systems, whereas silver's unparalleled conductivity makes it ideal for minimal-resistance applications, despite its susceptibility to oxidation. Advanced materials like graphene and carbon nanotubes (CNTs) are emerging as potential game-changers. Their exceptional electrical properties, mechanical flexibility, and high thermal conductivity suggest a future of conductive films with unprecedented performance in high-voltage environments.

4.2 Insulating materials

Insulating materials play a crucial role in maintaining safety and performance by preventing unintended current flow and ensuring dielectric integrity. Silicon dioxide (SiO_2) is a staple material due to its excellent insulating properties and compatibility with silicon-based technologies. Thin films of SiO_2 are extensively used in gate dielectrics and passivation layers. Silicon nitride (Si_3N_4) offers high dielectric strength and thermal stability, making it a preferred choice in capacitors and other high-voltage applications. Aluminum oxide (Al_2O_3) combines high dielectric strength with superior thermal properties, ensuring reliable insulation in power electronics. Emerging materials such as boron nitride (BN) and diamond-like carbon (DLC) are expanding the boundaries of insulation technology. These advanced materials provide high thermal conductivity alongside robust dielectric properties, addressing the challenges of thermal management in high-power devices.

4.3 Semiconducting materials

Semiconducting materials are the foundation of power electronic devices, enabling the efficient control and conversion of electrical energy. Silicon (Si) remains the most widely used material, offering well-understood properties, high availability, and cost-effectiveness. However, the demand for higher efficiency and performance has driven the adoption of advanced WBG semiconductors like gallium nitride (GaN) and silicon carbide (SiC). GaN excels in high-power and high-frequency applications due to its superior efficiency, high electron mobility, and elevated breakdown voltage. It is particularly popular in power conversion and RF amplifier systems. SiC, with its remarkable thermal conductivity and ability to operate under high-temperature conditions, is transforming applications such as electric vehicle inverters and power supplies. Recent advances in materials like zinc oxide (ZnO) and 2D materials such as transition metal dichalcogenides (TMDs) are providing new opportunities for thin film semiconductors. These materials exhibit unique electronic properties that can be leveraged for improved device performance and new functionalities.

4.4 Comparison of material properties

A comparative analysis of the properties of conductive, insulating, and semiconducting materials highlights their suitability for various high-voltage and power electronics applications. Conductive materials like copper and graphene offer superior electrical conductivity, while insulating materials such as Si_3N_4 and Al_2O_3 provide high dielectric strength and thermal stability. Semiconducting materials like

GaN and SiC enable efficient power conversion and high-temperature operation. Understanding the properties and applications of these materials enables the design and development of high-performance thin film devices for high-voltage and power electronics, driving technological advancements and efficiency improvements.

5. Thin films in capacitors

Thin films play a pivotal role in the evolution of capacitors, enhancing their performance, efficiency, and reliability in various applications, especially in high-voltage and power electronics. Capacitors, essential for energy storage, filtering, and signal processing in electronic circuits, benefit greatly from thin film technology due to their ability to deliver superior dielectric properties and high capacitance per unit area while supporting device miniaturization.

Thin films function as the dielectric layers in capacitors, separating conductive plates to store electric charge when a voltage is applied. The dielectric constant of the material used directly affects the capacitance value: higher dielectric constants enable the design of capacitors with higher capacitance in smaller volumes. This is critical for modern electronics, where space-saving and efficiency are paramount. Moreover, thin films provide better electrical insulation, reduced leakage currents, and improved thermal stability compared to bulk materials, making them indispensable in advanced electronic systems.

5.1 Role of thin films in capacitor performance

Thin films enhance capacitor performance in several ways. They enable compact designs by providing high capacitance densities, which is crucial for applications where size is a limiting factor. Thin film capacitors are particularly beneficial in high-voltage and power systems, where they support efficient energy transfer and storage. The precision of thin film deposition techniques ensures uniformity and consistency across the dielectric layer, minimizing defects and improving long-term reliability. Additionally, thin films allow for tailored material properties, such as dielectric strength and thermal stability, to meet the demands of specific applications, from portable devices to industrial power systems.

5.2 Types of thin films used in capacitors

Thin film capacitors utilize a variety of materials, each selected based on their specific performance characteristics. Silicon dioxide (SiO_2) is a prevalent choice due to its exceptional insulating properties and compatibility with silicon technologies, making it indispensable in metal-oxide-semiconductor (MOS) capacitors and other integrated devices. Titanium dioxide (TiO_2), with its high dielectric constant, is often employed in applications requiring substantial capacitance, such as dynamic random-access memory (DRAM), where its properties support high-performance and compact designs. Barium strontium titanate (BST), a ferroelectric material known for its high dielectric constant, finds use in tunable capacitors and microwave-frequency devices, excelling in scenarios requiring adaptability to varying electrical fields. Hafnium oxide (HfO_2), recognized for its high dielectric strength and thermal stability, is widely utilized in high-power and high-temperature environments, including automotive systems and energy storage solutions. Aluminum oxide (Al_2O_3), valued for its balance of dielectric strength and thermal performance, is a key material in medical and aerospace applications, where reliability and durability are paramount. These diverse materials enable thin film capacitors to meet the varied demands of modern electronic and power systems.

5.3 Advantages and challenges

Thin films enable capacitors with high capacitance per unit area, supporting the miniaturization of electronic components while maintaining high energy density. They exhibit excellent thermal and electrical stability, making them ideal for demanding environments like high-voltage and high-power systems. Advanced deposition techniques, such as atomic layer deposition (ALD) and sputtering, ensure precise control over the film's thickness and composition, resulting in consistent performance. Furthermore, thin film capacitors can be seamlessly integrated into semiconductor devices, improving functionality and reducing overall system size.

Despite their advantages, thin film capacitors face challenges in manufacturing and material selection. Sophisticated deposition techniques, while effective, are costly and require complex equipment, which can limit scalability. Additionally, achieving high dielectric constants alongside stability under high-voltage conditions is a persistent challenge in material development. Heat management in thin film capacitors is another critical concern; poor thermal dissipation can lead to performance degradation and reduced lifespan, especially in high-power applications.

5.4 Case studies and applications

Thin film capacitors have been successfully implemented in various high-voltage and power electronic applications, demonstrating their versatility and effectiveness.

i. In electric vehicles (EVs), thin film capacitors are used in power inverters and onboard chargers. The high capacitance density and thermal stability of materials like HfO_2 enable these capacitors to manage the high-power demands and temperature variations in EV systems. This contributes to improved energy efficiency and reliability.

ii. Medical devices such as implantable cardioverter defibrillators (ICDs) and external defibrillators rely on capacitors with high energy density and reliability. Thin film capacitors made from materials like TiO_2 and Al_2O_3 meet these stringent requirements, ensuring that these devices function accurately under critical conditions. Their compact design and thermal stability make them ideal for life-saving applications where performance consistency is non-negotiable.

6. Thin films in resistors

6.1 Enhancing resistor performance with thin films

Thin film resistors are integral components in electronic circuits, offering precise resistance values, stability, and reliability. The application of thin films in resistors enhances their performance by providing superior temperature coefficients, lower noise, and improved long-term stability. The thin film deposition process allows for fine control over the thickness and composition of the resistive material, enabling the production of resistors with highly accurate and consistent resistance values.

6.2 Material selection for thin film resistors

The choice of materials for thin film resistors is critical in defining their electrical and thermal performance, with different materials tailored to specific applications. Nickel-chromium (NiCr) is one of the most used materials due to its stable resistive properties and favorable temperature coefficient of resistance (TCR). NiCr provides a balanced combination of conductivity and resistance, making it versatile for various applications. Tantalum nitride (TaN) is another popular choice, renowned for its exceptional stability and corrosion resistance, which are particularly valuable in harsh environments and high-reliability scenarios. Its low TCR further enhances its performance stability over a wide temperature range. Silicon-chromium (SiCr) thin films are preferred for high-precision applications, as they offer excellent stability and a low TCR. These properties make SiCr ideal for precision resistors in integrated circuits and other applications requiring consistent and reliable performance.

6.3 Fabrication techniques

The fabrication of thin film resistors involves several key steps, including substrate preparation, film deposition, patterning, and passivation. The choice of fabrication technique significantly influences the performance and characteristics of the resulting resistors.

The substrate, typically made of materials like alumina, silicon, or glass, is cleaned and prepared to ensure good adhesion to the thin film. Surface treatments may include chemical cleaning, plasma etching, and deposition of adhesion layers.

Thin films are deposited using techniques such as sputtering, evaporation, and ALD. Sputtering is one of the most used methods due to its ability to produce uniform and adherent films. The deposition parameters, such as temperature, pressure, and gas composition, are carefully controlled to achieve the desired film properties.

Photolithography and etching processes are used to define the resistor geometry. The patterning process involves coating the substrate with a photoresist, exposing it to UV light through a mask, and developing the exposed photoresist to create the desired pattern.

6.4 Applications in high-voltage electronics

Thin film resistors find numerous applications in high-voltage and power electronics due to their precise resistance values, stability, and reliability. Some of the key applications include:

i. Thin film resistors are used in voltage dividers to provide accurate voltage levels in high-voltage circuits. Their stability ensures consistent performance over time and varying environmental conditions.

ii. Precision thin film resistors are employed in current sensing applications to measure current flow accurately. Their low TCR and high stability make them ideal for use in high-precision measurement systems.

7. Introduction to wide bandgap materials

Wide bandgap (WBG) materials are semiconductors with a larger bandgap than traditional silicon, typically greater than 2.3 eV. This characteristic enables them to operate at higher voltages, temperatures, and frequencies, making them ideal for high-voltage and power electronic applications. The primary WBG materials include silicon carbide (SiC), gallium nitride (GaN), and diamond. These materials are increasingly being integrated into thin films to enhance the performance and efficiency of electronic devices.

WBG materials have unique properties that distinguish them from conventional semiconductors. For example, SiC has a bandgap of 3.3 eV, GaN has a bandgap of 3.4 eV, and diamond boasts an even wider bandgap of 5.5 eV. These properties allow WBG materials to withstand higher electric fields, conduct current more efficiently, and operate at elevated temperatures without degradation.

7.1 Advantages of traditional insulators

WBG materials offer several advantages over traditional insulators like silicon dioxide (SiO_2) and silicon nitride (Si_3N_4):

i. WBG materials can withstand significantly higher electric fields before breaking down, making them ideal for high-voltage applications. For instance, SiC can endure electric fields up to ten times greater than silicon.

ii. Materials like SiC and diamond exhibit excellent thermal conductivity, which is crucial for dissipating heat in high-power devices. This property helps in maintaining device performance and longevity under high-power conditions.

iii. The higher electron mobility in WBG materials allows for faster switching speeds and reduced power losses, enhancing the overall efficiency of electronic devices. GaN, for instance, enables the development of high-frequency transistors with minimal energy loss.

7.2 Integration of wide bandgap materials with thin films

Integrating WBG materials with thin films involves advanced deposition techniques to create high-quality, uniform layers. Common methods include:

i. CVD is widely used for depositing WBG materials like SiC and diamond. It allows for precise control over the film thickness and composition, which is essential for achieving the desired electrical and thermal properties.

ii. ALD is used for creating ultra-thin and uniform layers of WBG materials. This method is particularly advantageous for producing gate dielectrics in high-frequency transistors.

Integrating WBG materials with thin films enhances the performance of various electronic devices. For instance, SiC thin films are used in power MOSFETs (metal-oxide-semiconductor field-effect transistors), GaN thin films in high electron mobility transistors (HEMTs), and diamond films in heat spreaders and high-frequency devices.

7.3 Case studies and practical applications

SiC power MOSFETs are increasingly being used in electric vehicles (EVs) due to their high efficiency and ability to operate at higher temperatures. The interest in SiC thin films in these devices results in lower conduction and switching losses, enhancing the overall efficiency and range of EVs. Moreover, the excellent thermal conductivity of SiC helps in effective heat management, reducing the need for bulky cooling systems.

GaN HEMTs are widely used in RF and microwave applications due to their high electron mobility and efficiency at high frequencies. The integration of GaN thin films in these devices enables the development of compact and high-performance RF amplifiers, which are essential for telecommunications and radar systems. The high breakdown voltage of GaN also allows for higher power managing capabilities, making them suitable for both commercial and military applications.

8. Importance of insulation in high-voltage systems

Insulation is a critical component in high-voltage systems to ensure safety, reliability, and efficiency. Proper insulation prevents electrical discharge, reduces leakage currents, and protects components from damage due to high electric fields. The use of thin films for insulation in high-voltage applications offers several advantages, including the ability to produce uniform, defect-free layers with excellent electrical properties.

Thin film insulators are particularly beneficial in applications requiring compact designs and high performances, such as power electronics, where space constraints and thermal management are critical. They also offer enhanced performance in terms of dielectric strength and breakdown voltage, which are essential for maintaining system integrity under high-voltage conditions.

8.1 Thin film insulating materials

Several materials are used for thin film insulation in high-voltage applications. Among them, polyamide ZnO nanocomposites have garnered significant attention due to their superior electrical and mechanical properties.

Polyamide (PA), known for its excellent mechanical strength, thermal stability, and dielectric properties, serves as a robust matrix for nanocomposites.

ZnO nanoparticles are incorporated into the polyamide matrix to enhance its electrical properties. ZnO offers high dielectric strength and excellent thermal conductivity, making it ideal for high-voltage insulation.

The combination of polyamide and ZnO nanoparticles results in a nanocomposite with improved dielectric properties, high breakdown voltage, and better thermal management compared to traditional insulating materials.

8.2 Fabrication and application techniques

The fabrication of polyamide ZnO nanocomposite thin films involves several steps to ensure uniform dispersion of nanoparticles and optimal electrical properties.

ZnO nanoparticles are dispersed in a polyamide solution using ultrasonic agitation to ensure a uniform distribution. This step is critical to achieving consistent electrical properties throughout the film. The mixed solution is applied to the substrate using spin coating, a process that involves rotating the substrate at high speeds to spread the solution evenly. This technique allows for precise control over film thickness and uniformity.

8.3 Performance metrics and reliability

The performance and reliability of high-voltage insulating thin films are evaluated based on several metrics, including dielectric strength, breakdown voltage, thermal conductivity, and long-term stability.

8.3.1 Dielectric strength

The ability of the material to withstand electric fields without breaking down. Polyamide ZnO nanocomposites exhibit higher dielectric strength compared to traditional materials, ensuring better insulation performance.

8.3.2 Breakdown voltage

The maximum voltage the material can manage before electrical failure occurs. Nanocomposite films have been shown to withstand higher voltages, making them suitable for demanding high-voltage applications.

8.3.3 Thermal conductivity

Efficient heat dissipation is crucial to prevent thermal breakdown. The inclusion of ZnO nanoparticles significantly enhances the thermal conductivity of polyamide films, improving their performance under high-power conditions.

8.3.4 Long-term stability

The material's ability to maintain its properties over time under operational stresses. Polyamide ZnO nanocomposite films have demonstrated excellent long-term stability, making them dependable for extended use in high-voltage systems.

In summary, the use of polyamide ZnO nanocomposite thin films offers significant advantages for high-voltage insulation. Their superior dielectric properties, enhanced thermal conductivity, and long-term stability make them ideal for advanced high-voltage and power electronic applications.

9. Miniaturization and efficiency in power electronics

9.1 Role of thin films in device miniaturization

Thin films play a crucial role in the miniaturization of power electronic devices, allowing for the reduction in size without compromising performance. By leveraging the unique properties of thin film materials, such as high surface area to volume ratio and superior electrical characteristics, manufacturers can create compact devices with enhanced functionality.

Thin films enable the integration of multiple components onto a single substrate, thereby reducing the overall footprint of the device. For example, thin film capacitors and resistors can be incorporated directly into integrated circuits (ICs), eliminating the need for bulky discrete components. This integration is particularly important in applications like portable electronics and aerospace systems, where space and weight are critical considerations.

9.2 Improving efficiency with thin films

Thin films contribute significantly to the efficiency of power electronic devices. Thin films can reduce energy losses and enhance thermal management by optimizing the material properties and deposition processes. Thin film transistors, such as those made from gallium nitride (GaN) and silicon carbide (SiC), exhibit lower on-resistance and faster switching speeds compared to traditional silicon-based devices. This results in reduced conduction and switching losses, thereby improving overall efficiency.

The use of high-k dielectric materials in thin film capacitors reduces leakage currents and allows for higher capacitance values in a smaller footprint. Materials like hafnium oxide (HfO_2) and barium strontium titanate (BST) are commonly used for this purpose.

9.3 Future trends and technologies

The future of power electronics will continue to be shaped by advances in thin film technologies. Key trends include:

The exploration of 2D materials, such as graphene and transition metal dichalcogenides (TMDs), is expected to revolutionize thin film applications. These materials offer exceptional electrical, thermal, and mechanical properties, which could lead to the development of even more compact and efficient devices.

The use of nanostructured thin films, such as quantum dots and nanowires, is being explored to enhance the performance of power electronic devices. Efficient transparent conducting oxides (TCOs) can provide superior electrical and optical properties, opening new possibilities for high-efficiency energy conversion and storage.

The integration of these advanced thin film technologies will lead to continued improvements in the miniaturization and efficiency of power electronics, enabling new applications and enhancing existing ones.

10. Practical applications and case studies

10.1 Real-world applications in power electronics

Thin films are integral to numerous real-world applications in power electronics, enhancing device performance, reliability, and efficiency. These applications span across various industries, including automotive, aerospace, renewable energy, and consumer electronics.

Thin films are used in power inverters and onboard chargers in EVs to improve efficiency and thermal management. High-performance thin film capacitors and resistors ensure reliable operation under high-voltage and temperature conditions.

Solar inverters and wind turbine controllers benefit from thin film technology, which enables higher efficiency and better heat dissipation. Thin film transistors and diodes are used to manage the conversion and distribution of power generated from renewable sources.

High-voltage and power electronic systems in aerospace applications rely on thin films for their superior performance in extreme environments. Thin films provide the necessary insulation and thermal management for avionics and satellite systems.

10.2 Case studies of thin film implementations

Gallium nitride (GaN) thin films are widely used in power amplifiers for telecommunications due to their high electron mobility and thermal conductivity. A leading telecommunications company implemented GaN-based amplifiers in their 5G base stations, resulting in a 30% increase in power efficiency and a significant reduction in cooling requirements. This advancement not only improved the performance but also reduced operational costs.

A prominent electric vehicle manufacturer integrated silicon carbide (SiC) MOSFETs with thin film technology into their inverters. The SiC MOSFETs provided higher switching speeds and lower conduction losses compared to traditional silicon devices. This integration led to a 20% improvement in the overall efficiency of the powertrain and extended the driving range of the vehicle by 15%.

10.3 Challenges and solutions

Despite the numerous advantages, the implementation of thin films in high-voltage and power electronics faces several challenges:

Ensuring compatibility between thin films and substrates can be challenging, particularly for new materials like GaN and SiC. Advanced deposition techniques and surface treatments are required to enhance adhesion and minimize defects. Managing heat dissipation in thin films is critical, especially in high-power applications. Incorporating materials with high thermal conductivity, such as diamond-like carbon (DLC), can address this issue.

10.4 Lessons learned from industry applications

Comprehensive testing and integration processes are crucial for ensuring the reliability of thin film components in high-voltage and power electronics. Rigorous testing under real-world conditions helps identify potential issues early and improve design robustness. Through practical applications and case studies, it is evident that thin films are integral to the advancement of various industries. However, addressing the challenges and learning from industry experiences are crucial for continued progress and innovation in this field.

11. Future directions and innovations

11.1 Emerging trends in thin film technology

The future of thin film technology in high-voltage and power electronics is marked by dynamic advancements and innovative trends that promise to redefine performance standards. Emerging materials and fabrication techniques are driving these transformations, with new applications further expanding the possibilities. Two-dimensional materials, such as graphene and transition metal dichalcogenides (TMDs), are at the forefront due to their exceptional electrical, thermal, and mechanical properties. Graphene's high conductivity and flexibility make it a strong candidate for use in flexible electronics and high-performance transistors. Meanwhile, nanostructured thin films, incorporating elements like quantum dots and nanowires, are enhancing electrical and optical properties, paving the way for high-efficiency energy conversion and storage devices. Additionally, smart materials capable of responding to environmental stimuli such as temperature, light, or electric fields are finding roles in adaptive systems, autonomous technologies, and smart grids, indicating their growing importance in modern applications.

In high-voltage applications, innovations in thin film technology are addressing the increasing demands for enhanced performance, reliability, and efficiency. Advanced dielectric materials, such as barium titanate ($BaTiO_3$) and hafnium oxide (HfO_2), are enabling higher capacitance values and better energy storage capabilities, critical for the development of next-generation capacitors and power devices. Ultra-thin film transistors utilizing materials like gallium nitride (GaN) and silicon carbide (SiC) are delivering improvements in switching speeds and energy efficiency, making them essential for high-frequency and advanced power management applications. Furthermore, nanocomposite insulators, including polyamide ZnO composites, offer superior dielectric properties and mechanical robustness, ensuring improved reliability and performance in extreme conditions. These advancements collectively highlight the transformative potential of thin films in addressing future challenges and enabling cutting-edge solutions in high-voltage and power electronics.

Developing precise and scalable deposition techniques, such as atomic layer deposition (ALD) and molecular beam epitaxy (MBE), will be critical for producing high-quality thin films. These techniques enable the fabrication of uniform and defect-free films necessary for high-performance devices.

Rigorous testing and reliability assessments are necessary to ensure that thin film components can withstand the demands of high-voltage applications. Research into accelerated life testing and failure analysis will help improve the durability and performance of these components.

12. Conclusion

12.1 Summary of key points

This chapter has comprehensively examined the critical role of thin films in high-voltage and power electronics, highlighting their properties, deposition methods, materials, applications, and future prospects. Thin films stand out for their unique electrical, thermal, and mechanical properties, which make them indispensable in high-voltage applications. Their high surface-to-volume ratio and ability to form uniform, defect-free layers provide distinct advantages in modern electronic devices.

Advanced deposition techniques such as Atomic Layer Deposition (ALD), Chemical Vapor Deposition (CVD), and Molecular Beam Epitaxy (MBE) allow precise control over film thickness and composition. These methods enable the production of high-quality thin films tailored for specific applications, ensuring performance optimization. Materials discussed include wide-bandgap semiconductors like GaN and SiC, high-k dielectrics such as HfO_2, and nanocomposites like polyamide ZnO, each offering remarkable properties that enhance the efficiency and reliability of power electronic devices.

Thin films play an integral role in the miniaturization and performance improvement of components such as capacitors, resistors, and transistors. Their application in real-world technologies, including electric vehicles, renewable energy systems, and telecommunications, underscores their growing importance in addressing modern technological challenges. Despite their potential, the field still faces challenges related to material compatibility, thermal management, and scaling up manufacturing processes. However, emerging innovations, such as flexible electronics, nanostructured films, and smart materials, hold promise for overcoming these obstacles and paving the way for future advancements in thin film technology.

12.2 Impact of thin films on high-voltage and power electronics

Thin films have had a profound impact on the field of high-voltage and power electronics, driving advancements in device performance, efficiency, and miniaturization. Their ability to enhance electrical insulation, reduce energy losses, and improve thermal management has led to the development of more compact and reliable power electronic systems. In electric vehicles, for instance, thin film technology has enabled the creation of more efficient powertrain components, extended vehicle range, and reduced energy consumption.

The integration of wide bandgap materials such as GaN and SiC into thin films has revolutionized power transistors and diodes, enabling faster switching speeds, higher breakdown voltages, and improved thermal performance. These advancements are crucial for applications requiring high-power density and efficiency, such as 5G telecommunications and aerospace systems.

12.3 Future outlook and potential developments

The future of thin films in high-voltage and power electronics holds immense promise, with numerous opportunities for innovation and expansion into new applications. One of the most exciting prospects lies in the continued advancement of novel materials. Research into two-dimensional materials such as graphene and transition metal dichalcogenides (TMDs) is expected to revolutionize thin film technology. These materials exhibit exceptional electrical, thermal, and mechanical properties, making them ideal candidates for next-generation electronic devices with unprecedented performance capabilities.

Another significant area of development is in flexible and stretchable electronics. Thin films designed for flexibility and elasticity will enable breakthroughs in wearable technology, medical devices, and flexible displays. Advances in organic thin film transistors (OTFTs) and flexible substrates are set to drive this trend, opening new possibilities for lightweight, portable, and adaptable electronics.

Nanostructured thin films are also at the forefront of innovation. By incorporating quantum dots, nanowires, and other nanoscale structures, the performance of thin films in energy conversion and storage devices can be greatly enhanced. These advancements will contribute to the creation of high-efficiency solar cells, long-lasting batteries, and ultra-capacitors, meeting the increasing demand for sustainable energy solutions.

The integration of smart thin films represents another transformative development. These materials, capable of responding to environmental stimuli such as temperature, light, or electric fields, are poised to play a crucial role in advanced sensing technologies and adaptive systems. Smart thin films will be instrumental in the evolution of autonomous systems and the implementation of intelligent energy grids.

In conclusion, thin films are at the forefront of technological advancements in high-voltage and power electronics, offering unparalleled benefits in terms of performance, efficiency, and miniaturization. By leveraging ongoing research and development efforts, the potential for future innovations in this field is vast, promising continued improvements and new applications that will shape the future of electronic devices and systems.

Author details

Nafees Ahmad
U.S.-Pakistan Center for Advanced Studies in Energy, National University of Sciences and Technology, Islamabad, Pakistan

*Address all correspondence to: sheikhnafees888@gmail.com

References

[1] Dhirendra Prasad R, Teli SB, Prasad RS, Prasad RB, Prasad SR, Sinha P, et al. A review on thin film technology and nanomaterial characterization techniques. ES Materials & Manufacturing. 2024;**25**:1198. DOI: 10.30919/esmm1198

[2] Yu Y, Gong M, Dong C, Xu X. Thin-film deposition techniques in surface and interface engineering of solid-state lithium batteries. Next Nanotechnology. 2023;**3-4**:100028. DOI: 10.1016/j.nxnano.2023.100028

[3] He J, Zhao T, Jing X, Demerdash NAO. Application of wide bandgap devices in renewable energy systems - benefits and challenges. In: 2014 International Conference on Renewable Energy Research and Application (ICRERA). Milwaukee, WI, USA; 2014. pp. 749-754. DOI: 10.1109/ICRERA.2014.7016485

[4] Yuvaraja S, Khandelwal V, Tang X, Li X. Wide bandgap semiconductor-based integrated circuits. Chip. 2023;**2**(4):100072. DOI: 10.1016/j.chip.2023.100072

[5] Adhikari P, Ghassemi M. A comprehensive review of mitigation strategies to address insulation challenges within high voltage, high power density (U)WBG power module packages. IEEE Transactions on Dielectrics and Electrical Insulation. 2024. DOI: 10.1109/TDEI.2024.3382070

[6] Wang Q, Chen G. Effect of nanofillers on the dielectric properties of epoxy nanocomposites. Advances in Materials Research. 2012;**1**:93-107. DOI: 10.12989/amr.2012.1.1.093

[7] Gharehghani A, Rabiei M, Mehranfar S, Saeedipour S, Mahmoudzadeh Andwari A, García A, et al. Progress in battery thermal management systems technologies for electric vehicles. Renewable and Sustainable Energy Reviews. 2024;**202**:114654. DOI: 10.1016/j.rser.2024.114654

[8] Johnson RW, Hultqvist A, Bent SF. A brief review of atomic layer deposition: From fundamentals to applications. Materials Today. 2014;**17**(5):236-246. DOI: 10.1016/j.mattod.2014.04.026

[9] Singh M, Tiwary SK, Nejat R, Douglas JF, Karim A. Anomalously high dielectric strength and capacitive energy density of thin entangled glassy polymer films. JACS Au. 2025;**5**(1):121-135. DOI: 10.1021/jacsau.4c00833

[10] Vishwas M, Rao K, Gowda KV, Chakradhar R. Optical, electrical and dielectric properties of TiO_2-SiO_2 films prepared by a cost effective sol-gel process. Spectrochimica Acta Part A, Molecular and Biomolecular Spectroscopy. 2011;**83**:614-617. DOI: 10.1016/j.saa.2011.08.009

[11] Abu-Thabit NY, Makhlouf ASH. Fundamental of smart coatings and thin films: Synthesis, deposition methods, and industrial applications. In: Advances in Smart Coatings and Thin Films for Future Industrial and Biomedical Engineering Applications. Elsevier; 2019. pp. 3-35. DOI: 10.1016/B978-0-12-849870-5.00001-X

[12] Doll G, Mensah B, Mohseni H, Scharf TW. Chemical vapor deposition and atomic layer deposition of coatings for mechanical applications. Journal of Thermal Spray Technology. 2009;**19**:510-516. DOI: 10.1007/s11666-009-9364-8

[13] Hezel R, Jaeger-Hezel K. Silicon nitride and aluminum oxide—multifunctional dielectric layers crucial for the progress of silicon solar cells. In: Petrova-Koch V, Hezel R, Goetzberger A, editors. High-Efficient Low-Cost Photovoltaics. Springer Series in Optical Sciences. Vol. 140. Cham: Springer; 2020. DOI: 10.1007/978-3-030-22864-4_4

Chapter 5

Functional Transition Metal Oxides: Novel Photocatalytic Applications and Engineering Perspectives

Celal Yelgel and Övgü C. Yelgel

Abstract

Solar-powered catalytic reactions are attractive for addressing energy and environmental issues. Photocatalysts are chemically stable semiconductors that initiate chemical reactions in adsorbed molecules when photoexcited charge carriers diffuse to the surface. The interface properties of photocatalysts play a crucial role in improving charge separation and charge transfer processes in photocatalytic reactions. Transition metal oxides (TMOs) are widely used as photocatalysts due to their excellent stability, affordability, abundance on Earth, and remarkable catalytic activity. Recent advancements in nanomaterial growth have significantly accelerated research in this field. This allows for fine-tuning catalyst performance by controlling the morphology of powder catalysts. Recent developments in surface characterisation of oxides have provided a deeper understanding of the atomic structure of surfaces and their interaction with the environment. The discovery of novel surface phases exhibiting unique electrical characteristics has opened up new routes for enhancing the performance of photocatalytic materials.

Keywords: transition metal oxides, mixed metal oxides, 2D materials, functional materials, photocatalysis

1. Introduction

The world is currently facing two major challenges: energy shortage and environmental contamination. These issues are primarily caused by rapid industrial development, population growth, and the excessive consumption of fossil fuels such as coal, oil, and natural gas [1]. As a result, greenhouse gas emissions are increasing, and hazardous wastes are being discharged. Thus, the advancement of highly efficient, eco-friendly technologies for producing clean energy and addressing environmental issues becomes a crucial undertaking. Sunlight is an abundant and readily accessible natural energy source that has significant promise in facilitating environmentally friendly photochemical processes. The process of converting solar energy into chemical energy, often known as solar fuels, is widely regarded as a very promising long-term solution for addressing global energy and environmental challenges [1, 2]. The use of a photocatalyst, often referred to as artificial photosynthesis, is essential in achieving this conversion. As shown in **Figure 1**, photocatalysis is the process of speeding up a reaction using a light-sensitive catalyst, similar to how plants utilise chlorophyll to transform water and carbon dioxide into oxygen and glucose with the help of sunlight [3]. The concept of photocatalysis centres around employing a photocatalyst to enhance chemical reactions when exposed to light. Upon absorbing light that meets or exceeds its bandgap energy, the photocatalyst facilitates the excitation of electrons from the valence band to the conduction band, resulting in the formation of electron-hole pairs. The charge carriers move towards the surface of the catalyst, where they trigger oxidation and reduction reactions with nearby molecules. The openings facilitate the oxidation of substances such as water or organic pollutants, while the electrons engage in the reduction of species like oxygen, resulting in the creation of reactive intermediates that propel the reaction forward. This method finds extensive application in areas like pollutant breakdown, water dissociation, and energy transformation.

Photooxidation: The surface of the photocatalyst includes water, referred to as "absorbed water." This water undergoes oxidation via positive holes generated in the valence band, caused by the transition of electrons to the conduction band owing to light irradiation, hence facilitating the creation of hydroxyl (OH•) radicals, which possess significant oxidative decomposition capability. Subsequently, these hydroxyl radicals interact with the organic substances included in the colours. In the presence of oxygen during this process, the intermediate radicals in the organic compounds, together with the oxygen molecules, may undergo radical chain reactions, potentially depleting oxygen in some instances. In such instances, organic matter eventually decomposes into carbon dioxide and water. In such conditions, organic

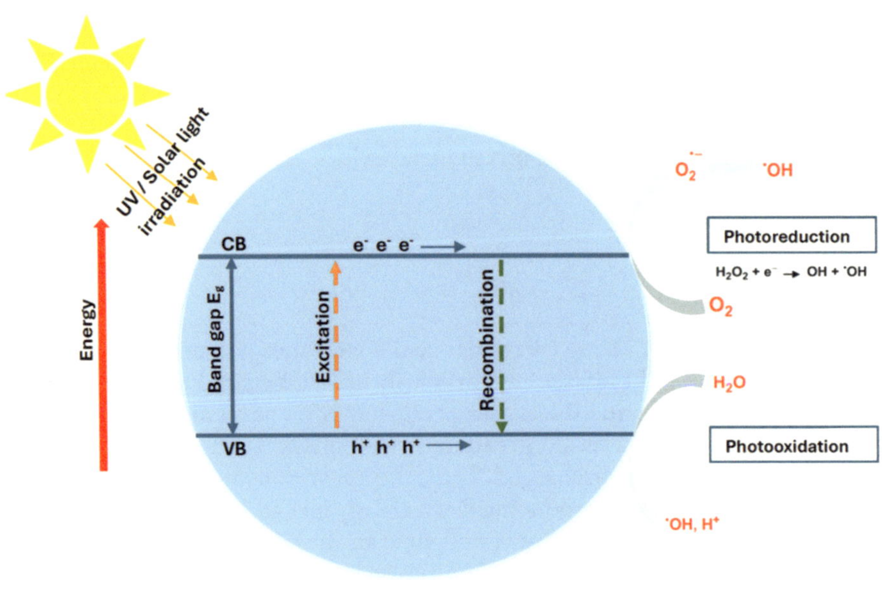

Figure 1.
A schematic diagram demonstrating the principle of photocatalysis.

substances might directly react with positive holes, leading to oxidative breakdown. Photoreduction: The reduction of oxygen occurs as an alternative to hydrogen production since oxygen is a readily reducible material. Electrons in the conduction band interact with dissolved oxygen species to generate superoxide anions. Superoxide anions bind to intermediate products in the oxidative process, resulting in the formation of peroxide or conversion to hydrogen peroxide and subsequently to water. The decrease is expected to transpire more readily in organic materials than in water. Consequently, a greater concentration of biological materials often elevates the quantity of positive holes. This reduces carrier recombination and amplifies photocatalytic activity. Important challenges in the field of photocatalysis involve addressing several key challenges. Firstly, it is crucial to tuning the band gap of the catalyst to enhance the photoelectric effect within the visible wavelength range. Secondly, efficient separation of photoelectrons and their corresponding holes is necessary to prevent recombination. Lastly, aligning the energy levels of the catalyst and reactant is essential for facilitating a redox reaction using the photoelectrons and holes. In order to efficiently drive the water splitting reaction using photocatalysis, it is crucial for the catalyst to have a conduction band minimum (CBM) that is higher than the potential required for dihydrogen evolution (H_2/H_2O): $H^+ \rightarrow H_2$ at around 0 eV versus normal hydrogen electrode (NHE), while the valence band maximum (VBM) needs to be lower than the potential of dioxygen evolution (O_2/H_2O): $O^{2-} \rightarrow O_2$ at around 1.23 eV versus NHE [4–8].

Typical photocatalytic materials consist of metal-free organic compounds and semiconductors [9–12]. Among these, semiconductors have received the most attention from researchers. A suitable light source with energy equal to or higher than the semiconductor's bandgap energy excites electrons from the valence band (VB) to the conduction band (CB), resulting in the formation of holes in the valence band. Some of the electrons and holes generated by light make their way to the surface of the semiconductor, while the rest combine and release energy in the form of heat or photons [9]. Redox reactions with absorbed species on the semiconductor surface can be caused by photoexcitons, depending on their chemical potential. As an illustration, when electrons interact with water molecules on the semiconductor surface, they can generate H_2 gas, while holes can produce O_2 below a certain potential [13]. When carbon dioxide gas is present with water, the electrons produced by light may reduce the carbon dioxide to form carbon monoxide, methane, or other carbohydrates. This process competes with the generation of hydrogen due to similar redox potentials. There are potential reactions that can occur in the VB below OH^-/H_2O potential, where holes may interact with surface-bound H_2O or OH^- to produce hydroxyl radicals. At the same time, electrons in the CB are utilised by electron-accepting species to generate radical anions [14]. Photocatalytic oxidation processes eliminate organic compounds and induce oxidative stress to microbial cell membranes, utilising hydroxyl radicals and radical anions as the primary oxidising agents. In recent years, there has been significant focus on exploring various semiconductors in different forms such as thin films, nanoparticles, and one-dimensional nanostructures [15–17]. In the field of photocatalysis, these different forms of semiconductors garnered considerable attention and effort from scientists and researchers. Functional transition metal oxides are a category of inorganic compounds that consist of transition metals and oxygen, specifically those elements located in the d-block of the periodic table. The diverse range of physical and chemical properties of these materials is extensively researched, allowing for customization to meet specific application needs. Their designation as "functional" stems from the impressive adaptability they demonstrate in their structural, electronic, magnetic, and optical characteristics, frequently reacting to external influences like temperature, pressure, or electric and magnetic fields. Transition metal oxides (TMOs) are very significant materials that have a broad variety of uses in sectors such as chemistry, energy storage, and electronics. TMOs, such as TiO_2, $SrTiO_3$, and ZnO., have been extensively investigated for their ability to catalyse chemical reactions using light [12]. TMOs are widely regarded as the most popular type of photocatalytic semiconductors due to their tunable structure, strong durability, affordability, abundance in the Earth's crust, and attractive catalytic properties [18]. Additionally, TMOs have a wide range of diverse compounds within this category. The utilisation of TMOs as photocatalysts is highly advanced from an engineering perspective and holds significant economic importance. Significant advancements have been achieved in recent times in the field of photocatalysis. These advancements have been made possible by the emergence of sophisticated experimental and theoretical techniques, enabling researchers to delve into the intricate chemistry occurring at the interface between transition metal oxide and the fluid and gaseous phase. Various TMOs have been extensively studied and used in the literature for a wide range of photocatalytic processes. These include Cu_2O, ZnO, NiO, Mn_3O_4, Co_3O_4, Fe_2O_3, CeO_2, ZrO_2, and TiO_2 [19–29].

The discovery of graphene has sparked a significant amount of research interest in other 2D materials, with the goal of achieving novel functionalisation and applications [30]. The exceptional physical and molecular properties of this atomically thin carbon layer with honeycomb lattices, which was initially achieved by exfoliating graphite crystals, have piqued significant research interest [31–33]. To investigate novel electronic, optical, or energy applications, substantial endeavours have been undertaken to investigate alternative 2D materials, including such as h-BN, TMOs, transition metal dichalcogenides (TMDs), silicene, germanene, MXcenes, phosphorene, metal oxide monolayers, group IV–VI or I–V monochalcogenides monolayers, and borophene [34–48]. Many of the different types of 2D materials are actually derived from their parent van der Waals crystals. As a result, their atomic structures closely resemble those of the bulk counterparts, with only slight differences in lattice constants due to the reduced dimensionality and absence of interlayer interaction. Some 2D monolayers, such as silicene, originate from bulk materials with covalent bonds [37]. While the bonding between atoms may undergo significant changes when transitioning from the bulk material to a 2D monolayer, the atomic structures remain similar. Nevertheless, a small atomic relaxation may take place to maintain a balanced bonding coordination in the 2D structures. Photocatalysts based on two-dimensional (2D) TMOs have gained considerable interest in tackling global issues such as energy scarcity and environmental pollution. 2D TMOs have shown remarkable surface area and unique physiochemical, electronic, and optical properties, which make them ideal for various photocatalytic applications.

This chapter provides a comprehensive analysis of TMOs, including experimental and theoretical aspects. It discusses various synthetic processes, characterisation methods, theoretical modelling approaches, and potential applications in photocatalysis. Various structures can be observed in TMOs, ranging from basic oxides to complex mixed-metal oxides and 2D nanostructured materials. The wide range of structures allows for precise manipulation of surface characteristics such as morphology, porosity, and surface area, all of which play a crucial role in determining photocatalytic performance. Finally, the various challenges, prospects, and future research directions/perspectives for developing efficient 2D TMOs are also provided.

DOI: http://dx.doi.org/10.5772/intechopen.1008556

2. Utilisation of 2D TMOs in photocatalytic applications

The atomically thin 2D structures exhibit a variety of extraordinary physicochemical and electronic characteristics as a result of their ultrathin thickness and unique morphology. This enables them to outperform their bulk counterparts and develop unique properties that are suitable for a wide range of engineering applications. Atomically thin 2D structures exhibit a high surface-to-volume ratio, a large lateral dimension, and a high specific surface area, all of which are advantageous for light harvesting. A greater number of active sites in interfacial catalytic reactions is also provided by the profuse coordination-unsaturated surface atoms in atomically thin 2D structures. Reactivity is enhanced by the unsatisfied bonding situation at the surface, and the presence of suspended bonds at the margins enables further manipulation of physical and chemical reactivity. The electron confinement in 2D, the large surface-to-volume ratio, and the structural reconstructions confer remarkable physical and chemical properties on 2D TMOs. Various 2D TMOs have been successfully synthesised, and their properties have been shown in superlattices, nanocomposites, and multi-layered systems. The exceptional characteristics of planar TMOs have already led to several practical applications as listed in **Figure 2**. Moreover, the ability to dynamically rearrange these materials and their ability to adapt various binding configurations hold great potential for future advancements.

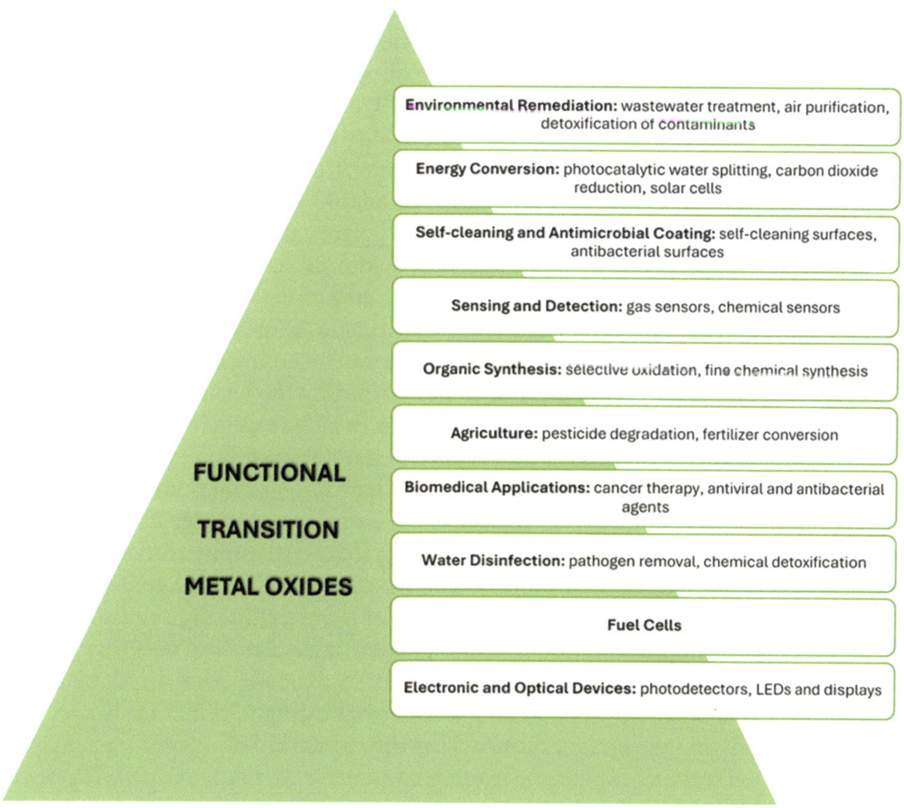

Figure 2.
Utilisation of TMO-based photocatalysts in various fields.

2.1 Structural features and synthesis of 2D TMOs

The TMOs composed of oxides of various metals exhibit notable variations when compared to their bulk counterparts. These 2D TMOs can take the form of single planar structures, multiple layers of individual oxides, or combinations of different oxides in superlattice configurations. The wide range of physical and chemical properties found in these materials allows for a diverse array of applications in fields such as optics, electronics, catalysis, energy storage, sensors, biosystems, piezoelectronics, thermal units, and superconductors [49–52]. The distinct nature of the oxygen ion plays a crucial role in producing extraordinary properties on the surface of a TMO. In the case of layered and 2D TMOs, these surface properties have a significant influence on the material's behaviour [53, 54]. The polarisability of O^{2-} ions is quite significant. The unique properties of planar TMOs arise from their ability to have large, nonlinear, and non-uniform distributions of charges within their lattices. This results in electrostatic screening on length scales of 1–100 nm, which gives rise to exceptional local surface and interfacial properties. Consequently, distinct energy states appear near or on the surface of planar TMOs, which differ greatly from their bulk counterparts. These surface states also lead to a strong attraction between 2D TMOs and neighbouring ions, known as Columbic interactions [55]. When two TMO surfaces are brought close together, interface potentials are created, causing a shift in the Fermi levels of the surface planes and significantly impacting their electronic properties. The electronic states in these planar TMOs may arise from various atomic orbitals. In addition, the ionic orbitals of TMO surfaces can be electronically activated through hybridisation. When the d orbitals of the 2D TMOs are arranged perpendicular to their planes, it can lead to the emergence of orbital and spin ordering, resulting in frustration [53, 54]. There is a vast selection of transition metals that can be used to create TMOs. The cationic species and their ability to change oxidation state play a crucial role in determining the fundamental properties of TMOs. In the field of 2D TMOs, a plethora of structures can be stabilised, thanks to the varying charge states and binding configurations of the cations [56, 57]. The various oxidation states of the metal result in a range of electronic properties, spanning from metallic to wide-gap insulating behaviour, depending on the stoichiometry of the TMO layer. The unique nature of the d-electronic states also leads to noticeable transitions between metal and insulator states under varying pressure and temperature conditions (such as Mott and Verwey transitions. A Mott transition is characterised by the localization of electrons as a result of strong electron-electron interactions, which results in an insulating state. A Verwey transition, on the other hand, is characterised by a structural and electronic transition that is frequently observed in mixed-valence compounds such as magnetite and is associated with changes in conductivity and charge ordering.). 2D TMOs exhibit a wide range of bandgaps, allowing for the manipulation of optical and electric properties across a broad spectrum of wavelengths [58]. The redox properties of nanostructured TMOs are unique, with many of them exhibiting reversible trends [59]. In addition, numerous 2D TMOs exhibit remarkable chemical and thermal stabilities [60]. The tunability of oxygen vacancies in nanostructured TMOs can be achieved through a range of established procedures, leading to remarkable alterations [61]. In addition, the energy required for the formation of oxygen vacancies in 2D TMOs may decrease significantly, leading to an increase in the reducibility of the metal cations.

Several TMOs, like MoO_3 [44, 62–64], WO_3 [65], Ga_2O_3 [66], and V_2O_5 [67, 68], can be naturally occurring as layered crystals in either hydrated or anhydrous forms.

The oxides can be exfoliated using liquid or gas phase techniques, resulting in nanosheets with stable oxygen-terminated basal surfaces that can withstand exposure to both air and water. Based on the composition of the unit cells in these TMOs, they can be reduced to a certain number of layers in their thinnest and most stable planar configurations. Several TMOs, including titanium and zinc oxides, lack a naturally occurring layered crystal phase. In such cases, the process typically involves the exfoliation of the materials from their more stable stratified compounds or their deposition using layer-by-layer techniques. Ensuring the stability of these structures after they have been grown is crucial and can be achieved by incorporating charged motifs or functional groups. It is important to consider that when synthesising these materials, the familiar bulk structures may transform into specific 2D arrangements. For example, the wurtzite structure can change into a planar configuration in 2D ZnO [69]. Perovskites belong to a vast group of 2D TMOs with the formula following a general pattern of ABO_3, with A-site ions positioned on the corners of the cubic unit cell and B-site ions, which are transition metal elements, located at the centre of the cubic unit cell. They can be synthesised into a 2D form using layer-by-layer deposition techniques, potentially followed by stabilisation steps. Layered perovskites are composed of ABO_3 layers that are separated by thin sheets containing motifs with A and B cations, which may have varying sizes. The characteristics of layered perovskites can vary based on the thickness of the motif layers and the lateral offsetting of the perovskite layers in relation to each other. The 2D plane cell defining thicknesses are determined by the value of n in the general formula of layered perovskites. There are three primary layered perovskites known as Aurivillius (AU), Dion-Jacobson (DJ), and Ruddlesden-Popper (RP) phases. If the bulk crystals of perovskites have a layered structure, similar to layered TMOs, they can also be easily separated into thin layers from their naturally occurring layered phases. Oxides that are sandwiched between insulators and metals, like planar cuprates and CoO_2, are typically referred to as quasi-2D TMOs [50, 70]. Due to the ability of electrons to move freely within these loosely connected oxide layers, these materials may exhibit superconducting properties. Layers of lanthanum, barium, and strontium ions are used to stabilise quasi-2D TMOs and introduce doping. Concisely, **Table 1** presents a comprehensive list of methodologies that have been provided and effectively used for the synthesis of 2D TMOs.

2D TMOs	Synthesis techniques	Reference
Cu_2O, Pd_5O_4, Ag_2O, TiO_2	Surface oxidation	[71–74]
FeO, CeO_2, Ti_2O_3, WO_3, $CuWO_4$, $FeWO_3$	Physical vapour deposition (PVD)	[75–81]
Y_2O_3	Chemical vapour deposition (CVD)	[82]
CuO, Eu_2O_3, Bi_2WO_6, MnO_2, Fe_3O_4	Self-assembly	[81, 83–85]
Co_3O_4	Pyrolysis	[86]
CeO_2, WO_3	Hydrothermal method	[87, 88]
SnO_2, ZnO	Solvothermal method	[89, 90]
MnO_2	In situ replacement	[91]
V_2O_5	Mechanical force-assisted liquid exfoliation	[68]

Table 1.
Synthesis techniques for several 2D transition metal oxides (TMOs) presented in the literature.

2.2 Exploring the theoretical aspects of 2D TMOs

The adhesion and contact of reactants with the surface, as well as the impact of surfaces on the adhesion and interactions between reactants, are crucial elements that greatly contribute to catalytic reactions. The TMO surface has several intricacies that offer challenges for both theoretical modelling and practical characterisation. When characterising surfaces and studying surface reactions, it is crucial to consider various descriptors. These include the stoichiometric ratios of cations and anions, the nature of covalent and ionic bonding, the charges of adsorbed species, the acidity and basicity of the surface, the presence of cation and anion vacancies, the mobility of oxygen, and the surface reconstruction caused by dangling bonds. The intricate nature of TMOs makes them one of the most intricate systems for both experimental and theoretical study. Theoretical computations enable the exploration of 2D TMOs without the need for prior synthesis. Typically, the examination of 2D TMOs as photocatalysts includes the use of numerous computational approaches. First-principles calculations using density functional theory (DFT) have emerged as a compelling and efficient method for studying 2D TMOs [92]. It can be used not only to explain experimental results, but also to anticipate novel 2D TMOs and analyse their related physical or chemical characteristics. Using DFT, one can optimise geometry and obtain electrical characteristics such as band structure, density of states (DOS), and charge distribution. Using hybrid functionals like HSE06 can greatly improve the precision of estimating band gaps. Time-dependent density functional theory (TD-DFT) is a powerful tool in the field of computational physics, allowing researchers to calculate various optical properties, including absorption spectra. In addition, the analysis of interactions with reactants such as water or CO_2 involves studying adsorption energies and reaction routes. Understanding the behaviour of charge carriers, including their mass and speed, is crucial for gaining insights into the separation of electrons and holes. The determination of catalytic activity involves analysing various characteristics, including the adsorption Gibbs free energy (ΔG) and the alignment of redox potential with water splitting processes. Investigating defects and the behaviour of excitons helps us gain insights into how charges are trapped and how carriers behave. On the other hand, conducting molecular dynamics (MD) simulations allows us to assess the thermodynamic stability. Utilising machine learning and high-throughput screening techniques can significantly accelerate the process of material discovery. By analysing large datasets obtained from DFT, researchers can make predictions about the photocatalytic characteristics of materials and explore novel 2D TMOs [93].

3. 2D TMOs as photocatalysts

Photocatalytic water splitting processes are being studied for their potential to provide clean and sustainable hydrogen. The procedure entails the dissociation of water into its fundamental components (H_2 and O_2) using light and a photocatalyst, usually a semiconductor. For optimal water splitting efficiency, the semiconductor's valence band must exceed the water oxidation potential (O_2/H_2O, 1.23 eV), while the conduction band must be lower than the hydrogen evolution potential (H^+/H_2, 0 eV). Furthermore, the photocatalyst must possess an optimal band gap above 1.23 eV to enable effective photon absorption [94, 95]. TMOs have undergone significant investigation as potential photocatalysts over the last few decades, attributed to their excellent chemical stability, affordability, and non-toxic nature. Specifically, the significant surface-to-volume ratio and the capacity to achieve adequate photon flux density for various electron reactions (such as water splitting: $2H_2O + 4e^- \rightarrow 2H_2 + O_2$) position 2D TMOs as more advantageous compared to 0D, 1D, and 3D TMOS. Reducing the thickness of 2D materials to the nanometre scale can enhance the quantum confinement effect, thereby enabling the manipulation of the optical and electronic properties of the catalyst to align with the wavelength of the light source and the energy level required for specific catalytic reactions. Moreover, 2D materials possess a distinctive surface architecture with a majority of atoms exposed, and the surface modification of these materials may enhance the catalyst's capacity for light absorption, molecule adsorption, and activation.

Notwithstanding the exceptional performance observed experimentally, the photocatalytic application of 2D TMOs remains impeded by various limitations, including extensive optical bandgaps, inadequate electronic conductivity (specifically elevated electron-hole recombination rates), and unsuitable band edge positions for reduction or oxidation reactions. Numerous strategies exist to enhance the photocatalytic activity of 2D TMOs [96]. Metal doping, nonmetal doping, and the introduction of oxygen vacancies have been documented as methods to lower the broad bandgaps of 2D TMOs for visible light harvesting by generating gap states [97–102]. Elemental doping is a traditional and efficacious method for altering the surface characteristics and band structures of bulk TMOs to optimise visible light absorption and boost redox activity [10]. The potential enhancement of mechanical flexibility in some 2D TMOs may provide superior integration of size-mismatched dopant ions inside the crystal structure compared to their bulk counterparts [103]. Dopants may be broadly categorised into metallic and nonmetallic elements [10]. Doping TMOs with nonmetal elements often results in a reduction of the bandgap, attributed to the overlap between the orbitals of the dopant and the oxygen atoms [104]. The development of localised states inside the bandgap by the dopant is also claimed to contribute to bandgap narrowing [105]. Thereby, the photogenerated electrons may be directly stimulated from the dopant gap state to the conduction band of the material, rather than from the valence band edge. The ionic radius and valence state of the dopant, in relation to the host metal cation, significantly influence the selection of occupied sites and the electronic characteristics of the dopant in metal doping. They may affect the surface characteristics, the placement of the Fermi level, and the conductivity inside the matrix [106]. Both metallic and nonmetallic dopants may enhance the separation of photogenerated charge carriers. Specifically, at suitable concentrations, the dopants may function as deep trap sites for one kind of charge carrier, while permitting another to access the material surface for intended redox reactions [10]. Nonmetal elements, including nitrogen, boron, fluorine, phosphorus, and carbon, have been examined for doping 2D TMOs. Liu et al. reported that the N dopant in MoS_2 enhances its photocatalytic capabilities, resulting in a RhB dye photodegradation (which denotes the process of decomposing Rhodamine B dye into smaller, less detrimental molecules by light exposure, often facilitated by catalysts such as photocatalytic materials) rate of 134 μmol $(g\ h)^{-1}$, while 2D MoS_2 nanosheets exhibit no performance [107]. Wang et al. investigated N-doped 2D TiO_2 by a standard hydrothermal method utilising nitric acid (HNO_3) as the nitrogen dopant source [108]. They have found that the H_2 generation rate for the N-doped nanosheets reaches 17.2 mmol $(g\ h)^{-1}$, which is about four and two times more than that of N-doped microcrystals and 2D bare nanosheets, respectively [109, 110]. The use of tetrafluoroboric acid (HBF_4) as the hydrothermal reaction agent during synthesis facilitates the co-doping of boron and fluorine in 2D TiO_2 nanosheets, which predominantly feature

{001} exposed facets and exhibit a visible light absorption edge at around 516 nm [108]. The dopants are essential for achieving a 2D planar nanostructure [111]. The introduction of oxygen vacancies in TMOs modifies the bandgap energy by creating localised defect states inside the bandgap, in addition to elemental doping. The electrical conductivity is simultaneously enhanced, which may decrease the charge recombination rate [112]. Moreover, the partial reduction of the metal oxidation state might serve as an electron trap, thus promoting the separation of photogenerated charges [112]. Surface plasmon resonance is thus stimulated in the visible or near-infrared spectrum, significantly enhancing the solar light harvesting efficiency of the material [63]. Pioneering studies on solo 2D MoO_{3-x} nanosheets and core-shell structured 2D MoO_{3-x}/MoO_3 nanosheets have shown better visible light absorption and dramatically higher photocatalytic activity compared to their stoichiometric counterparts [113, 114].

Depending on the specific photocatalytic application, the functionalisation of 2D TMOs with a particular organic group may enhance the adsorption of surface species, hence improving photocatalytic performance. Xue et al. functionalise 2D ZnO nanosheets with amine to augment the surface adsorption of CO_2 molecules for photocatalytic CO_2 reduction. Monoethanolamine (MEA) is used for its hydroxyl group, facilitating covalent attachment to ZnO, and its primary amine group, which imparts an amine-functionalized surface. The effective formation of C−N bonds with CO_2 in terminal amine groups is thought to activate CO_2 and yield carbamate. The presence of carbamate may facilitate direct interactions with Zn_2^+ to acquire electrons from ZnO and initiate reduction processes involving CO and CH_4. The experiment demonstrated that the generation rates of both CO and CH_4 from amine-functionalized ZnO nanosheets are double those of their bare counterparts [115]. Moreover, inspired by dye-sensitised solar cells, the sensitisation of dye molecules on semiconductor surfaces is a novel method to improve visible light absorption [116, 117]. Incorporating 2D TMO systems may result in a greater quantity of adsorbed dyes owing to the extensive surface area provided by the 2D planar design [118]. Noble metal-free zinc porphyrin complexes have been extensively researched as photosensitisers for dye-sensitised solar cells, achieving a maximum conversion efficiency of 10%. Zinc porphyrin has significant absorption in the visible light spectrum, exceptional photochemical stability, and favourable redox potentials for electron injection and dye renewal, making it an appealing candidate for photosensitisers due to its delocalised p electrons [119]. Recent results indicate the incorporation of zinc porphyrin into 2D MoS_2 nanosheet heterostructures with TiO_2 and ZnO [118, 119]. Besides the restricted efficacy of visible light absorption, the rapid recombination of photogenerated charge carriers in several individual 2D TMOs is a significant issue. Developing a 2D TMOs-based heterojunction with an appropriately chosen semiconductor may mitigate this issue [10]. An optimal alignment of CB and VB energy levels between TMOs and the semiconductor may provide an efficient transfer channel for photogenerated charge carriers between the two materials. The predominant method is type II band alignment. The photogenerated electrons migrate from the more positive conduction band edge to the less positive one, while the holes move from the more negative valence band edge to the less negative one, therefore achieving spatial charge carrier separation [120]. Nonetheless, several parameters, including defect density and crystallinity, may substantially modify the band topologies of the materials, hence affecting the coupling efficiency [10]. Moreover, the disparity in dimensions and size of linked semiconductors may be significant for hetero-interfacial connections [121]. In contrast to its 2D/low-dimensional equivalents, the 2D/2D heterostructure has enhanced stability and coupling at heterointerfaces, attributable to its extensive contact surface and reduced exciton diffusion length, which promote the transfer and separation of photoexciton pairs [121]. The heterojunctions of 2D MoS_2 with low-dimensional TMOs have been extensively studied include TiO_2, MoO_3, and ZnO [122–128]. In addition to type II band alignment, the establishment of p-n heterojunctions is an excellent method for mitigating charge recombination [129]. This method involves the interaction between a p-type material and its n-type counterpart, resulting in the formation of a space charge layer that generates an internal electric field, hence enhancing the likelihood of electron-hole separation. In addition to these studies, we can summarise the important findings reported in the current literature on this subject as follows: a typical example of TMO photocatalysts is TiO_2 nanosheet. Han et al. discovered that (001)-terminated TiO_2 nanosheets have superior photocatalytic activity [130]. It is significant that, in addition to extensively studied photocatalytic water splitting, 2D TMOs are emerging as potential photocatalysts for several value-added chemical processes in recent years [131, 132]. Chen et al. synthesised ultrathin, single-crystal WO_3 nanosheets using 2D orientated attachment [131]. The researchers discovered that dimensionality reduction modifies the bandgap of WO_3, enabling the WO_3 nanosheet to facilitate the photocatalytic reduction of CO_2 to CH_4 in the presence of water. Li et al. found that CO may be generated by the photocatalytic oxidation of CH_4 on anatase (101)-terminated TiO_2, achieving an impressive yield of 81.9% [132]. Their proposal indicates that the increased selectivity for CO production arises from the presence of surface Ti_3^+ sites. Zhou et al. have developed a bottom-up approach for the creation of monolayer materials exhibiting rapid charge separation and a highly active surface [84]. They synthesised Bi_2WO_6 nanosheets of monolayer thickness and a sandwich substructure of $[BiO]^+$-$[WO_4]^{2-}$-$[BiO]^+$, in which the sub-layers are interconnected by robust chemical bonds rather than weak van der Waals interactions. The presence of unsaturated Bi atoms results in a highly active surface and a reduced bandgap compared to Bi_2WO_6 nanocrystals. Upon irradiation, ultrafast charge transfer and separation were observed in this heterostructure, specifically Bi_2WO_6, with photogenerated holes on the surface and photogenerated electrons in the intermediate layer. This phenomenon resulted in enhanced photocatalytic performance for the visible-light-driven degradation of Rhodamine B and hydrogen production.

4. Summary and future perspectives

The use of 2D TMOs in photocatalysts has gained significant traction in recent years. The characteristics of 2D TMOs exhibit notable benefits for photocatalytic hydrogen evolution reaction, pollutant degradation, carbon reduction, and microbial disinfection processes. The extensive surface area of 2D nanostructures and their 3D hierarchical derivatives provide more reaction sites for photocatalytic processes and enhance charge transfer, hence enhancing the rates of photocatalytic reactions. The distinctive adjustable band structures in 2D TMOs result in more thermodynamically advantageous redox reactions during photocatalytic processes. Moreover, the principal exposed crystal facets afforded by the 2D geometric design significantly enhance the photocatalytic efficacy. The exploration of 2D TMOs-based photocatalysts remains nascent in comparison to the more developed 1D and nanoparticle-based photocatalytic systems. Numerous chances are now arising, and substantial progress is anticipated in the near future. The essential factor for advancing the study domain is the efficient and high-yield synthesis of 2D TMOs, particularly those with non-layered crystal structures. Further optimisation of deposition parameters for vapour-phase synthesis, including the quantity of source material and vaporisation energy/temperature, is necessary to maximise production yield suited for effective photocatalytic reactions. The potential application of innovative templates and growth directing agents in liquid-phase synthesis techniques presents significant opportunities for the large-scale production of 2D TMOs with robust morphological control, which is crucial for the advancement of efficient 2D TMOs-based photocatalysts that meet industry standards. This book chapter critically examines ways for enhancing photocatalytic performance, specifically emphasising elemental doping, surface functionalisation, and the creation of heterojunctions with both semiconducting and conductive materials. In contrast to conventional 2D/low-dimensional heterojunctions, the 2D/2D system is posited to have enhanced stability and superior coupling at heterointerfaces, hence promoting photoinduced charge transfer and separation. Nevertheless, the associated synergetic effects are now under-researched and need additional investigation for the development of novel 2D/2D heterojunctions exhibiting exceptional photocatalytic characteristics. Moreover, the adjustment of stoichiometry in 2D TMOs significantly affects their electronic band structures, leading to reduced bandgap energies and potentially inducing surface plasmon resonance in the visible light spectrum when the concentration of free charge carriers is sufficiently high. The employment of the sub-stoichiometric phase of 2D TMOs in photocatalytic processes may provide many advantages, such as enhanced light harvesting and improved transfer of photoexcited charge carriers. To summarise, there are several opportunities for novel discoveries in 2D TMOs-based photocatalysts, necessitating substantial resources and focus to uncover new phenomena that use the distinctive advantages of 2D materials.

Acknowledgements

C. Yelgel and Ö. C. Yelgel acknowledge the support from the University of Manchester, National Graphene Institute and School of Physics and Astronomy.

Conflict of interest

The authors declare no conflict of interest.

DOI: http://dx.doi.org/10.5772/intechopen.1008556

Author details

Celal Yelgel[1,2] and Övgü C. Yelgel[2,3]*

1 Department of Electricity and Energy, Recep Tayyip Erdoğan University, Rize, Turkey

2 The Computational Science and Machine Learning Laboratory, Recep Tayyip Erdoğan University, Rize, Turkey

3 Department of Electrical-Electronics Engineering, Recep Tayyip Erdoğan University, Rize, Turkey

*Address all correspondence to: oceyda.yelgel@erdogan.edu.tr

References

[1] Zou Z, Ye J, Sayama K, Arakawa H. Direct splitting of water under visible light irradiation with an oxide semiconductor photocatalyst. Nature. 2001;**414**(6864):625

[2] Bahnemann D. Photocatalytic water treatment: Solar energy applications. Solar Energy. 2004;**77**(5):445

[3] Kapilashrami M, Zhang Y, Liu YS, Hagfeldt A, Guo J. Probing the optical property and electronic structure of TiO_2 nanomaterials for renewable energy applications. Chemical Reviews. 2014;**114**:9662

[4] Fujishima A, Honda K. Electrochemical photolysis of water at a semiconductor electrode. Nature. 1972;**238**:37

[5] Maeda K, Domen K. New non-oxide photocatalysts designed for overall water splitting under visible light. Journal of Physical Chemistry C. 2007;**111**:7851

[6] Himpsel FJ, Cook PL, de la Torre G, Garcia-Lastra JM, Gonzalez-Moreno R, Guo JH, et al. Design of solar cell materials via soft X-ray spectroscopy. Journal of Electron Spectroscopy and Related Phenomena. 2013;**190**(Part A):2

[7] Linsebigler AL, Lu G, Yates JT. Photocatalysis on TiO_2 surfaces: Principles, mechanisms, and selected results. Chemical Reviews. 1995;**95**:735

[8] Jing D, Guo L, Liang Z, Zhang X, Liu H, Li M, et al. Efficient solar hydrogen production by photocatalytic water splitting: From fundamental study to pilot demonstration. International Journal of Hydrogen Energy. 2010;**35**:7087

[9] Li W, Li L, Xiao H, Qi R, Huang Y, Xie Z, et al. Iodo-BODIPY: A visible-light-driven, highly efficient and photostable metal-free organic photocatalyst. RSC Advances. 2013;**3**(32):13417

[10] Zhang H, Chen G, Bahnemann DW. Photoelectrocatalytic materials for environmental applications. Journal of Materials Chemistry. 2009;**19**(29):5089

[11] Maeda K, Domen K. Photocatalytic water splitting: Recent progress and future challenges. Journal of Physical Chemistry Letters. 2010;**1**(18):2655

[12] Hernández-Alonso MD, Fresno F, Suárez S, Coronado JM. Development of alternative photocatalysts to TiO_2: Challenges and opportunities. Energy & Environmental Science. 2009;**2**(12):1231

[13] Maeda K, Teramura K, Lu D, Takata T, Saito N, Inoue Y, et al. Photocatalyst releasing hydrogen from water. Nature. 2006;**440**(7082):295

[14] Turchi CS, Ollis DF. Photocatalytic degradation of organic water contaminants: Mechanisms involving hydroxyl radical attack. Journal of Catalysis. 1990;**122**(1):178

[15] Tong H, Ouyang S, Bi Y, Umezawa N, Oshikiri M, Ye J. Nano-photocatalytic materials: Possibilities and challenges. Advanced Materials. 2012;**24**(2):229

[16] Banerjee S, Mohapatra SK, Das PP, Misra M. Synthesis of coupled semiconductor by filling 1D TiO_2 nanotubes with CdS. Chemistry of Materials. 2008;**20**(21):6784

[17] Wu Y, Yan H, Yang P. Semiconductor nanowire array: Potential substrates for

photocatalysis and photovoltaics. Topics in Catalysis. 2002;**19**(2):197

[18] Dong YC, Ghuman KK, Popescu R, Duchesne PN, Zhou WJ, Loh JY, et al. Advancement of Science. 2018;**5**:1700732

[19] Sayed M, Zhang LY, Yu JG. Plasmon-induced interfacial charge-transfer transition prompts enhanced CO_2 photoreduction over Cu/Cu_2O octahedrons. Chemical Engineering Journal. 2020;**397**:125390

[20] Hermans Y, Klein A, Sarker HP, Huda MN, Junge H, Toupance T, et al. Pinning of the Fermi level in $CuFeO_2$ by Polaron formation limiting the photovoltage for photochemical water splitting. Advanced Functional Materials. 2020;**30**:1910432

[21] Deng HZ, Xu FY, Cheng B, Yu JG, Ho WK. Photocatalytic CO_2 reduction of C/ZnO nanofibers enhanced by an Ni-NiS cocatalyst. Nanoscale. 2020;**12**:7206

[22] Liu JN, Jia QH, Long JL, Wang XX, Gao ZW, Gu Q. Amorphous NiO as co-catalyst for enhanced visible-light-driven hydrogen generation over g-C_3N_4 photocatalyst. Applied Catalysis. B, Environmental. 2018;**222**:35

[23] Li N, Tian Y, Zhao JH, Zhang J, Zhang J, Zuo W, et al. Efficient removal of chromium from water by Mn_3O_4@ZnO/Mn_3O_4 composite under simulated sunlight irradiation: Synergy of photocatalytic reduction and adsorption. Applied Catalysis. B, Environmental. 2017;**214**:126

[24] Gao C, Meng QQ, Zhao K, Yin HJ, Wang DW, Guo J, et al. Co_3O_4 hexagonal platelets with controllable facets enabling highly efficient visible-light photocatalytic reduction of CO_2. Advanced Materials. 2016;**28**:6485

[25] Xu QL, Zhu BC, Jiang CJ, Cheng B, Yu JG. Constructing 2D/2D Fe_2O_3/g-C_3N_4 direct Z-scheme photocatalysts with enhanced H_2 generation performance. Solar RRL. 2018;**2**:1800006

[26] Ruan CY, Huang ZQ, Lin J, Li L, Liu XY, Tian M, et al. Synergy of the catalytic activation on Ni and the CeO_2–TiO_2/$Ce_2Ti_2O_7$ stoichiometric redox cycle for dramatically enhanced solar fuel production. Energy & Environmental Science. 2019;**12**:767

[27] Wang Q, Edalati K, Koganemaru Y, Nakamura SH, Watanabe M, Ishihara T, et al. Photocatalytic hydrogen generation on low-bandgap black zirconia (ZrO_2) produced by high-pressure torsion. Journal of Materials Chemistry A. 2020;**8**:3643

[28] Meng AY, Zhang LY, Cheng B, Yu JG. Dual Cocatalysts in TiO_2 Photocatalysis. Advanced Materials. 2019;**31**:1807660

[29] Liu C, Zhang QF, Hou WH, Zou ZG. 2D titanium/niobium metal oxide-based materials for photocatalytic application. Solar RRL. 2020;**4**:2000070

[30] Novoselov KS, Geim AK, Morozov SV, Jiang D, Zhang Y, Dubonos SV, et al. Electric field effect in atomically thin carbon films. Science. 2004;**306**:666

[31] Novoselov KS, Geim AK, Morozov S, Jiang D, Katsnelson M, Grigorieva I, et al. Two-dimensional gas of massless Dirac fermions in graphene. Nature. 2005;**438**:197

[32] Singh V, Joung D, Zhai L, Das S, Khondaker SI, Seal S. Graphene based materials: Past, present and future. Progress in Materials Science. 2011;**56**:1178

[33] Castro Neto AH, Guinea F, Peres NMR, Novoselov KS, Geim AK. The electronic properties of graphene. Reviews of Modern Physics. 2009;**81**:109

[34] Kubota Y, Watanabe K, Tsuda O, Taniguchi T. Deep ultraviolet light-emitting hexagonal boron nitride synthesized at atmospheric pressure. Science. 2007;**317**:932

[35] Mak KF, Lee C, Hone J, Shan J, Heinz TF. Atomically thin MoS_2: A new direct-gap semiconductor. Physical Review Letters. 2010;**105**:136805

[36] Kuc A, Zibouche N, Heine T. Influence of quantum confinement on the electronic structure of the transition metal sulfide TS_2. Physical Review B. 2011;**83**:245213

[37] Vogt P, De Padova P, Quaresima C, Avila J, Frantzeskakis E, Asensio MC, et al. Silicene: Compelling experimental evidence for graphenelike two-dimensional silicon. Physical Review Letters. 2012;**108**:155501

[38] Balendhran S, Walia S, Nili H, Sriram S, Bhaskaran M. Elemental analogues of graphene: Silicene, germanene, stanene, and phosphorene. Small. 2015;**11**:640

[39] Li L, Lu SZ, Pan J, Qin Z, Wang YQ, Wang Y, et al. Buckled germanene formation on Pt (111). Advanced Materials. 2014;**26**:4820

[40] Naguib M, Mochalin VN, Barsoum MW, Gogotsi Y. 25th anniversary article: MXenes: A new family of two-dimensional materials. Advanced Materials. 2014;**26**:992

[41] Rodin A, Carvalho A, Neto AC. Strain-induced gap modification in black phosphorus. Physical Review Letters. 2014;**112**:176801

[42] Liu H, Neal AT, Zhu Z, Luo Z, Xu Z, Tománek D, et al. An unexplored 2d semiconductor with a high hole mobility. ACS Nano. 2014;**8**:4033

[43] Osada M, Sasaki T. Exfoliated oxide nanosheets: New solution to nanoelectronics. Journal of Materials Chemistry. 2009;**19**:2503

[44] Kalantar-Zadeh K, Tang J, Wang M, Wang KL, Shailos A, Galatsis K, et al. Synthesis of nanometre-thick MoO_3 sheets. Nanoscale. 2010;**2**:429

[45] Zhuang HL, Hennig RG. Single-layer group-III monochalcogenide photocatalysts for water splitting. Chemistry of Materials. 2013;**25**:3232

[46] Qin G, Qin Z, Fang WZ, Zhang LC, Yue SY, Yan QB, et al. Diverse anisotropy of phonon transport in two-dimensional group IV–VI compounds: A comparative study. Nanoscale. 2016;**8**:11306

[47] Mannix AJ, Zhou XF, Kiraly B, Wood JD, Alducin D, Myers BD, et al. Synthesis of borophenes: Anisotropic, two-dimensional boron polymorphs. Science. 2015;**350**:1513

[48] Peng B, Zhang H, Shao H, Xu Y, Zhang R, Zhu H. The electronic, optical, and thermodynamic properties of borophene from first-principles calculations. Journal of Physical Chemistry C. 2016;**4**:3592

[49] Xu M, Liang T, Shi M, Chen H. Graphene-like two-dimensional materials. Chemical Reviews. 2013;**113**:3766

[50] Takada K, Sakurai H, Takayama-Muromachi E, Izumi F, Dilanian RA, Sasaki T. Superconductivity in two-dimensional CoO_2 layers. Nature. 2003;**422**:53

[51] Walia S, Balendhran S, Nili H, Zhuiykov S, Rosengarten G, Wang QH, et al. Transition metal oxides—Thermoelectric properties. Progress in Materials Science. 2013;**58**:1443

[52] Gupta MK, Lee JH, Lee KY, Kim SW. Two-dimension alvanadium-doped ZnO nanosheet-based flexible direct current nanogenerator. ACS Nano. 2013;**7**:8932

[53] Mannhart J, Schlom DG. Oxide interfaces—An opportunity for electronics. Science. 2010;**327**:1607

[54] Goniakowski J, Finocchi F, Noguera C. Polarity of oxide surfaces and nanostructures. Reports on Progress in Physics. 2008;**71**:01650

[55] Chimene D, Alge DL, Gaharwar AK. Two-dimensional nanomaterials forbiomedical applications: Emerging trends and future prospects. Advanced Materials. 2015;**27**:7261

[56] Surnev S, Ramsey MG, Netzer FP. Vanadium oxide surface studies. Progress in Surface Science. 2003;**73**:117

[57] Wu QH, Fortunelli A, Granozzi G. Preparation, characterisation and structure of Ti and Al ultrathin oxide films on metals. International Reviews in Physical Chemistry. 2009;**28**:517

[58] Sasaki T, Watanabe M. Semiconductor nanosheet crystallites of quasi-TiO_2 and their optical properties. The Journal of Physical Chemistry. B. 1997;**101**:10159

[59] Poizot P, Laruelle S, Grugeon S, Dupont L, Tarascon JM. Nano-sized transition-metal oxides as negative-electrode materials for lithium-ion batteries. Nature. 2000;**407**:496

[60] Fukuda K, Ebina Y, Shibata T, Aizawa T, Nakai I, Sasaki T. Unusual crystallization behaviors of anatase nanocrystallites from a molecularly thin titania nanosheet and its stackedforms: Increase in nucleation temperature and oriented growth. Journal of the American Chemical Society. 2007;**129**:202

[61] Ganduglia-Pirovano MV, Hofmann A, Sauer J. Oxygen vacancies intransition metal and rare earth oxides: Current state of understanding andremaining challenges. Surface Science Reports. 2007;**62**:219

[62] Balendhran S, Walia S, Nili H, Ou JZ, Zhuiykov S, Kaner RB, et al. Two-dimensional molybdenum trioxide and dichalcogenides. Advanced Functional Materials. 2013;**23**:3952

[63] Alsaif MMYA, Latham K, Field MR, Yao DD, Medehkar NV, Beane GA, et al. Tunable plasmon resonances in two-dimensional molybdenum oxide nanoflakes. Advanced Materials. 2014;**26**:3931

[64] Alsaif MMYA, Chrimes AF, Daeneke T, Balendhran S, Bellisario DO, Son Y, et al. High-performance field effect transistors using electronic inks of 2D molybdenum oxide nanoflakes. Advanced Functional Materials. 2016;**26**:91

[65] Kalantar-zadeh K, Vijayaraghavan A, Ham MH, Zheng H, Breedon M, Strano MS. Synthesis of atomically thin WO_3 sheets fromhydrated tungsten trioxide. Chemistry of Materials. 2010;**22**:5660

[66] Feng W, Wang X, Zhang J, Wang L, Zheng W, Hu P, et al. Synthesis of two-dimensional β-Ga_2O_3 nanosheets for high-performance solar blind photodetectors. Journal of Materials Chemistry C. 2014;**2**:3254

[67] Teran-Escobar G, Pampel J, Caicedo JM, Lira-Cantu M.

Low-temperature, solution-processed, layered V_2O_5 hydrate as the hole-transport layer forstable organic solar cells. Energy & Environmental Science. 2013;**6**:3088

[68] Rui X, Lu Z, Yu H, Yang D, Hng HH, Lim TM, et al. Ultrathin V_2O_5 nanosheet cathodes: Realizing ultra fast reversible lithium storage. Nanoscale. 2013;**5**:556

[69] Tusche C, Meyerheim HL, Kirschner J. Observation of depolarized ZnO(0001) monolayers: Formation of unreconstructed planar sheets. Physical Review Letters. 2007;**99**:026102

[70] Damascelli A, Hussain Z, Shen ZX. Angle-resolved photoemission studies of the cuprate superconductors. Reviews of Modern Physics. 2003;**75**:473

[71] Zhou G, Luo L, Li L, Ciston J, Stach EA, Yang JC. Step-edge-induced oxide growth during the oxidation of Cu surfaces. Physical Review Letters. 2012;**109**:235502

[72] Lundgren E, Kresse G, Klein C, Borg M, Andersen JN, De Santis M, et al. Two-dimensional oxide on Pd (111). Physical Review Letters. 2002;**88**:246103

[73] Rovida G, Pratesi F, Maglietta M, Ferroni E. Chemisorption of oxygen on the silver (111) surface. Surface Science. 1974;**43**:230

[74] Tao J, Luttrell T, Batzill M. A two-dimensional phase of TiO_2 with a reduced bandgap. Nature Chemistry. 2011;**3**:296

[75] Kim Y, Westphal C, Ynzunza R, Galloway H, Salmeron M, Van Hove M, et al. Interlayer interactions in epitaxial oxide growth: FeO on Pt (111). Physical Review B. 1997;**55**:R13448

[76] Ma L, Doudin N, Surnev S, Barcaro G, Sementa L, Fortunelli A, et al. Lattice strain defects in a ceria nanolayer. Journal of Physical Chemistry Letters. 2016;**7**:1303

[77] Barcaro G, Agnoli S, Sedona F, Rizzi GA, Fortunelli A, Granozzi G. Structure of reduced ultrathin TiO_x polar films on Pt(111). Journal of Physical Chemistry C. 2009;**113**:5721

[78] Doudin N, Kuhness D, Blatnik M, Barcaro G, Negreiros F, Sementa L, et al. Nanoscale domain structure and defects in a 2-D WO_3 layer on Pd(100). Journal of Physical Chemistry C. 2016;**120**:28682

[79] Denk M, Kuhness D, Wagner M, Surnev S, Negreiros FR, Sementa L, et al. Metal tungstates at the ultimate two dimensional limit: Fabrication of a $CuWO_4$ nanophase. ACS Nano. 2014;**8**:3947

[80] Pomp S, Kuhness D, Barcaro G, Sementa L, Mankad V, Fortunelli A, et al. Two-dimensional iron tungstate: A ternary oxide layer with honeycomb geometry. Journal of Physical Chemistry C. 2016;**120**:7629

[81] Liu J, Huang X, Li Y, Sulieman K, He X, Sun F. Self-assembled CuO monocrystalline nanoarchitectures with controlled dimensionality and morphology. Crystal Growth & Design. 2006;**6**:1690

[82] Addou R, Dahal A, Batzill M. Growth of a two-dimensional dielectric monolayer on quasi-freestanding graphene. Nature Nanotechnology. 2013;**8**:41

[83] Zhong Y, Yang Y, Ma Y, Yao J. Controlled synthesis of ultrathin lamellar Eu_2O_3 nanocrystals: Self-assembly of 1D nanowires to 2D nanosheets. Chemical Communications. 2013;**49**:10355

[84] Zhou Y, Zhang Y, Lin M, Long J, Zhang Z, Lin H, et al. Monolayered Bi_2WO_6 nanosheets mimicking heterojunction interface with open surfaces for photocatalysis. Nature Communications. 2015;**6**:8340

[85] Sun Z, Liao T, Dou Y, Hwang SM, Park MS, Jiang L, et al. Generalized self-assembly of scalable two-dimensional transition metal oxide nanosheets. Nature Communications. 2014;**5**:3813

[86] Jiang Y, Chen L, Zhang H, Zhang Q, Chen W, Zhu J, et al. Two-dimensional Co_3O_4 thin sheets assembled by 3D interconnected nanoflake array framework structures with enhanced supercapacitor performance derived from coordination complexes. Chemical Engineering Journal. 2016;**292**:1

[87] Sun Y, Liu Q, Gao S, Cheng H, Lei F, Sun Z, et al. Pits confined in ultrathin cerium (IV) oxide for studying catalytic centers in carbon monoxide oxidation. Nature Communications. 2013;**4**:2899

[88] Harale N, Dalavi D, Mali SS, Tarwal N, Vanalakar S, Rao V, et al. Single-step hydrothermally grown nanosheet-assembled tungsten oxide thin films for sensitive and selective NO_2 gas detection. Journal of Materials Science. 2018;**53**:6094

[89] Sun Y, Lei F, Gao S, Pan B, Zhou J, Xie Y. Atomically thin tin dioxide sheets for efficient catalytic oxidation of carbon monoxide. Angewandte Chemie, International Edition. 2013;**52**:10569

[90] Liu X, Sun Y, Yu M, Yin Y, Du B, Tang W, et al. Enhanced ethanol sensing properties of ultrathin ZnO nanosheets decorated with CuO nanoparticles. Sensors and Actuators B: Chemical. 2018;**255**:3384

[91] Zhao G, Li J, Jiang L, Dong H, Wang X, Hu W. Synthesizing MnO_2 nanosheets from graphene oxide templates for high performance pseudosupercapacitors. Chemical Science. 2012;**3**:433

[92] Wang L, Maxisch T, Ceder G. Oxidation energies of transition metal oxides within the GGA+U framework. Physical Review B. 2006;**73**:195107

[93] Xu J, Cao XM, Hu P. Perspective on computational reaction prediction using machine learning methods in heterogeneous catalysis. Physical Chemistry Chemical Physics. 2021;**23**:11155

[94] Lin S, Huang H, Ma T, Zhang Y. Photocatalytic oxygen evolution from water splitting. Advancement of Science. 2021;**8**(1):2002458

[95] Bie C, Wang L, Yu J. Challenges for photocatalytic overall water splitting. Chem. 2022;**8**(6):1567

[96] Haque F, Daeneke T, Kalantar-zadeh K, Ou JZ. Two-dimensional transition metal oxide and chalcogenide-based photocatalysts. Nano-Micro Letters. 2018;**10**:23

[97] Ida S, Ishihara T. Recent progress in two-dimensional oxide photocatalysts for water splitting. Journal of Physical Chemistry Letters. 2014;**5**:2533

[98] Luo B, Liu G, Wang L. Recent advances in 2D materials for photocatalysis. Nanoscale. 2016;**8**:6904

[99] Lei F, Sun Y, Liu K, Gao S, Liang L, Pan B, et al. Oxygen vacancies confined in ultrathin indium oxide porous sheets for promoted visible-light water splitting. Journal of the American Chemical Society. 2014;**136**:6826

[100] Jin YJ, Linghu J, Chai J, Chua CS, Wong LM, Feng YP, et al. Defect

evolution enhanced visible-light photocatalytic activity in nitrogen-doped anatase TiO_2 thin films. Journal of Physical Chemistry C. 2018;**122**:16600

[101] Tao J, Yang M, Chai J, Pan J, Feng Y, Wang S. Atomic N modified rutile TiO_2(110) surface layer with significant visible light photoactivity. Journal of Physical Chemistry C. 2014;**118**:994

[102] Chai J, Yang M, Chen Q, Pan J, Zhang Z, Feng Y, et al. Effects of nitrogen incorporation on the electronic structure of rutile-TiO_2. Journal of Applied Physics. 2011;**109**:023707

[103] Zhang H. Ultrathin two-dimensional nanomaterials. ACS Nano. 2015;**9**(10):9451

[104] Asahi R, Morikawa T, Ohwaki T, Aoki K, Taga Y. Visiblelight photocatalysis in nitrogen-doped titanium oxides. Science. 2001;**293**(5528):269

[105] Bilmes S, Mandelbaum P, Alvarez F, Victoria N. Surface, electronic structure of titanium dioxide photocatalysts. The Journal of Physical Chemistry. B. 2000;**104**(42):9851

[106] Serpone N, Lawless D, Disdier J, Herrmann JM. Spectroscopic, photoconductivity, and photocatalytic studies of TiO_2 colloids: Naked and with the lattice doped with Cr^{3+}, Fe^{3+}, and V^{5+} cations. Langmuir. 1994;**10**(3):643

[107] Liu P, Liu Y, Ye W, Ma J, Gao D. Flower-like n-doped MoS_2 for photocatalytic degradation of RhB by visible light irradiation. Nanotechnology. 2016;**27**(22):225403

[108] Wang B, Leung MKH, Lu XY, Chen SY. Synthesis and photocatalytic activity of boron and fluorine codoped TiO_2 nanosheets with reactive facets. Applied Energy. 2013;**112**(16):1190

[109] Xu H, Ouyang S, Li P, Kako T, Ye J. High-active anatase TiO_2 nanosheets exposed with 95% 100 facets toward efficient H_2 evolution and CO_2 photoreduction. ACS Applied Materials & Interfaces. 2013;**5**(4):1348

[110] Xiang Q, Yu J, Wang W, Jaroniec M. Nitrogen self-doped nanosized TiO_2 sheets with exposed 001 facets for enhanced visible-light photocatalytic activity. Chemical Communications. 2011;**47**(24):6906

[111] Liu Y, Xie S, Li H, Wang X. A highly efficient sunlight driven ZnO nanosheet photocatalyst: Synergetic effect of P-doping and MoS_2 atomic layer loading. ChemCatChem. 2014;**6**(9):2522

[112] Seifollahi Bazarjani M, Hojamberdiev M, Morita K, Zhu G, Cherkashinin G, Fasel C, et al. Visible light photocatalysis with c-$WO_{(3-x)}$/ $WO_{3x}H_2O$ nanoheterostructures in situ formed in mesoporous polycarbosilane-siloxane polymer. Journal of the American Chemical Society. 2013;**135**(11):4467

[113] Cheng H, Kamegawa T, Mori K, Yamashita H. Surfactantfree nonaqueous synthesis of plasmonic molybdenum oxide nanosheets with enhanced catalytic activity for hydrogen generation from ammonia borane under visible light. Angewandte Chemie, International Edition. 2014;**53**(11):2910

[114] Tan X, Wang L, Cheng C, Yan X, Shen B, Zhang J. Plasmonic MoO_{3-x}@ MoO_3 nanosheets for highly sensitive SERS detection through nanoshell-isolated electromagnetic enhancement. Chemical Communications. 2016;**52**(14):2893

[115] Liao Y, Hu Z, Gu Q, Xue C. Amine-functionalized ZnO nanosheets for efficient CO_2 capture

and photoreduction. Molecules. 2015;**20**(10):18847

[116] Cho Y, Choi W, Lee CH, Hyeon T, Lee HI. Visible lightinduced degradation of carbon tetrachloride on dye-sensitized TiO_2. Environmental Science & Technology. 2001;**35**(5):966-970

[117] Takanabe K, Kamata K, Wang X, Antonietti M, Kubota J, Domen K. Photocatalytic hydrogen evolution on dye-sensitized mesoporous carbon nitride photocatalyst with magnesium phthalocyanine. Physical Chemistry Chemical Physics. 2010;**12**(40):13020

[118] Yuan Y, Lu H, Ji Z, Zhong J, Ding M, Chen D, et al. Enhanced visible-light-induced hydrogen evolution from water in a noblemetal-free system catalyzed by ZnTCPP-MoS_2/TiO_2 assembly. Chemical Engineering Journal. 2015;**275**:8

[119] Yuan YJ, Tu JR, Ye ZJ, Lu HW, Ji ZG, Hu B, et al. Visiblelight-driven hydrogen production from water in a noble-metalfree system catalyzed by zinc porphyrin sensitized MoS_2/ZnO. Dyes and Pigments. 2015;**123**:285

[120] Jafari T, Moharreri E, Amin AS, Miao R, Song W, Suib SL. Photocatalytic water splitting-the untamed dream: A review of recent advances. Molecules. 2016;**21**(7):900

[121] Zhou M, Lou XWD, Xie Y. Two-dimensional nanosheets for photoelectrochemical water splitting: Possibilities and opportunities. Nano Today. 2013;**8**(6):598

[122] Zhou W, Yin Z, Du Y, Huang X, Zeng Z, Fan Z, et al. Synthesis of few-layer MoS_2 nanosheet-coated TiO_2 nanobelt heterostructures for enhanced photocatalytic activities. Small. 2013;**9**(1):140

[123] Xiang Q, Yu J, Jaroniec M. Synergetic effect of MoS_2 and graphene as cocatalysts for enhanced photocatalytic H_2 production activity of TiO_2 nanoparticles. Journal of the American Chemical Society. 2012;**134**(15):6575

[124] Shen M, Yan Z, Yang L, Du P, Zhang J, Xiang B. MoS_2 nanosheet/TiO_2 nanowire hybrid nanostructures for enhanced visible-light photocatalytic activities. Chemical Communications. 2014;**50**(97):15447

[125] Liu C, Wang L, Tang Y, Luo S, Liu Y, Zhang S, et al. Vertical single or few-layer MoS_2 nanosheets rooting into TiO_2 nanofibers for highly efficient photocatalytic hydrogen evolution. Applied Catalysis. B, Environmental. 2015;**164**:1

[126] Zhang X, Shao C, Li X, Miao F, Wang K, Lu N, et al. 3D MoS_2 nanosheet/TiO_2 nanofiber heterostructures with enhanced photocatalytic activity under UV irradiation. Journal of Alloys and Compounds. 2016;**686**:137

[127] Zhong M, Wei Z, Meng X, Wu F, Li J. From MoS_2 microspheres to α-MoO_3 nanoplates: Growth mechanism and photocatalytic activities. European Journal of Inorganic Chemistry. 2014;**2014**(20):3245

[128] Yuan YJ, Wang F, Hu B, Lu HW, Yu ZT, Zou ZG. Significant enhancement in photocatalytic hydrogen evolution from water using a MoS_2 nanosheet-coated ZnO heterostructure photocatalyst. Dalton Transactions. 2015;**44**(24):10997

[129] Ida S, Takashiba A, Koga S, Hagiwara H, Ishihara T. Potential gradient and photocatalytic activity of an ultrathin p–n junction surface prepared with two-dimensional semiconducting

nanocrystals. Journal of the American Chemical Society. 2014;**136**(5):1872

[130] Han X, Kuang Q, Jin M, Xie Z, Zheng L. Synthesis of titania nanosheets with a high percentage of exposed (001) facets and related photocatalytic properties. Journal of the American Chemical Society. 2009;**131**:3152

[131] Chen X, Zhou Y, Liu Q, Li Z, Liu J, Zou Z. Ultrathin, single-crystal WO_3 nanosheets by two-dimensional oriented attachment toward enhanced photocatalystic reduction of CO_2 into hydrocarbon fuels under visible light. ACS Applied Materials & Interfaces. 2012;**4**:3372

[132] Li W, He D, Hu G, Li X, Banerjee G, Li J, et al. Selective CO production by photoelectrochemical methane oxidation on TiO_2. ACS Central Science. 2018;**4**:631